Franz-Theo Suttmeier

Numerical solution of Variational Inequalities
by Adaptive Finite Elements

VIEWEG+TEUBNER RESEARCH

Advances in Numerical Mathematics

Herausgeber | Editors:

Prof. Dr. Dr. h.c. Hans Georg Bock
Prof. Dr. Dr. h.c. Wolfgang Hackbusch
Prof. Mitchell Luskin
Prof. Dr. Rolf Rannacher

Franz-Theo Suttmeier

Numerical solution of Variational Inequalities by Adaptive Finite Elements

VIEWEG+TEUBNER RESEARCH

Bibliographic information published by the Deutsche Nationalbibliothek
The Deutsche Nationalbibliothek lists this publication in the Deutsche Nationalbibliografie;
detailed bibliographic data are available in the Internet at http://dnb.d-nb.de.

1st Edition 2008

All rights reserved
© Vieweg+Teubner | GWV Fachverlage GmbH, Wiesbaden 2008

Readers: Christel A. Roß

Vieweg+Teubner is part of the specialist publishing group Springer Science+Business Media
www.viewegteubner.de

Cover design: KünkelLopka Medienentwicklung, Heidelberg
Printing company: Strauss Offsetdruck, Mörlenbach
Printed on acid-free paper
Printed in Germany

ISBN 978-3-8348-0664-2

Summary

This work describes a general approach to *a posteriori* error estimation and adaptive mesh design for finite element models where the solution is subjected to inequality constraints. This is an extension to *variational inequalities* of the so-called *Dual-Weighted-Residual* method (*DWR* method) which is based on a variational formulation of the problem and uses global *duality arguments* for deriving weighted *a posteriori* error estimates with respect to arbitrary functionals of the error. In these estimates local residuals of the computed solution are multiplied by sensitivity factors which are obtained from a numerically computed *dual* solution. The resulting local error indicators are used in a feed-back process for generating economical meshes which are tailored according to the particular goal of the computation. This method is developed here for several model problems. Based on these examples, a general concept is proposed, which provides a systematic way of adaptive error control for problems stated in form of *variational inequalities*.

Für Alexandra, Katharina und Merle

Contents

Chapter 1

Introduction

The work at hand is devoted to the numerical treatment of systems of partial differential equations, where the solution is subjected to inequality constraints. We employ the finite element Galerkin (FE) method to obtain approximate solutions of such systems, which for instance typically arise in the field of continuum mechanics. Examples are plastic materials, where certain norms of the stresses are bounded or contact problems, where the displacement is restricted by a rigid obstacle. For illustration, the situation of a workpiece pressed onto a grinding disk (Figures 1.1,1.2) is approximated by a FE-scheme (Figures 1.3,1.4) to approximate the resulting surface forces. The basis for applying an FE discretisation is a suitable mathematical setting, which in the topics under consideration takes the form of variational inequalities (VI).

We intend to develop efficient schemes in the sense that the requirements on computer storage and computation time are close to the minimum to achieve a prescribed accuracy, or under given computer resources the FE-approximation is nearly optimal with respect to the discretisation error. In order to reach this goal, we employ an adaptive algorithm, i.e., based on a suitable criterion locally refined FE meshes are produced within an iteration. The heart of each such procedure is an *a posteriori* error bound. On the one hand such an estimator provides information how to refine the meshes with respect to a given quantity of interest prescribed by the application with minimum computer resources. For example in grinding processes engineers are interested in the distribution of normal stresses in the contact zone between tool and material.

On the other hand it enables us to get a reliable quantification of the accuracy of our numerical approximation. The design of an abstract framework to provide *a posteriori* error bounds for FE-solutions of variational inequalities is the main focus of the present work.

Figure 1.1: Snap shot of a grinding process.

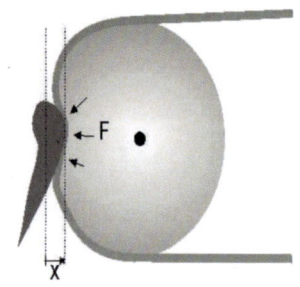

Figure 1.2: Sketch of the model situation.

Figure 1.3: FE-Simulation

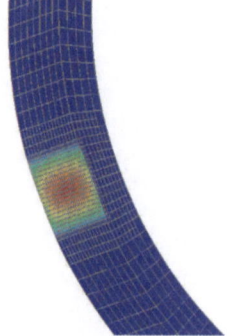

Figure 1.4: Zoom.

Solving complex systems of partial differential equations by discretisation methods may be considered in the context of *model reduction*: a conceptually *infinite* dimensional model is approximated by a *finite* one. Here, the quality of the approximation depends on a proper choice of the discretisation parameters, e.g., the mesh width, the polynomial degree of trial functions or the size of certain stabilisation parameters. As the result of the computation, we obtain an approximation to the desired output quantity of the simulation and besides that certain accuracy indicators like local cell-residuals. Controlling the error in such an approximation of a continuous model of a physical system requires to determine the influence factors of the *local* error indicators on the target quantity. Such a sensitivity analysis with respect to local perturbations of the model is common in optimal control theory and introduces the concept of a *dual* or *adjoint* problem.

For illustration consider a *continuous* model governed by a differential operator L and a forcing term f, and an approximating *discrete* model depending on a discretisation parameter $h > 0$:

$$Lu = f, \quad L_h u_h = f_h.$$

In controlling this discretisation, we have to detect the interplay of the various error propagation effects in order to achieve (i) *a posteriori error control*, i.e., control of the error in quantities of physical interest like point values of deflection and stress or mean stresses along parts of the boundary, etc., and (ii) *solution-adapted* meshing, i.e., design of economical meshes for computing these quantities with best efficiency. Our error analysis is based on the computable *residual* $\rho(u_h) = f - Lu_h$, which is well defined in the context of a Galerkin finite element method. Then the error $e := u - u_h$ satisfies the *error equation* $Le = \rho(u_h)$.

Traditionally, *a posteriori* error estimation in Galerkin finite element methods is done with respect to the natural *energy norm* $\|.\|_E$ induced by the underlying differential operator

$$\|u - u_h\|_E \leq c \sup_{\|\varphi\|_E = 1} |\langle \rho(u_h), \varphi \rangle|.$$

This approach was initiated by the pioneering work of Babuška and Rheinboldt and has been further developed by Ladevéze & Leguillon [47], Bank & Weiser [6], and Babuška & Miller [3]. A particularly simple and useful variant was introduced by Zienkiewicz and Zhu [74]. For discussions and more references see the survey articles by Ainsworth and Oden [1] and Verfürth [72]. This approach seems rather generic as it is directly based on the variational formulation of the problem and allows to exploit its natural coercivity. However, in most applications the error in the energy norm does not provide a useful bound on the error in the quantities of real physical interest. A more versatile method for *a posteriori* error estiamtion with respect to relevant error measures like the L^2-norm over subdomains, point values, line averages, etc. is obtained by using duality arguments as well known from the *a priori* error analysis of finite element methods (so-called Aubin-Nitsche trick). This approach is particularly designed for achieving high solution accuracy at minimum computational cost. The additional work required by the evaluation of the error bounds is acceptable since, particularly in nonlinear cases, it usually amounts to only a small fraction of the total cost.

Let $J(u)$ be a quantity of physical interest derived from the solution u by applying a functional $J(\cdot)$. The goal is to control the error $J(u) - J(u_h)$ of the discretisation with respect to this functional output in terms of the computable cell residuals $\rho_T(u_h)$. An example is control of the total error $e_T = (u - u_h)_{|T}$ in some mesh cell T. By superposition, e_T splits into two components, the locally produced *truncation error* and the globally transported *pollution error*, $e_T = e^{\mathrm{loc}}{}_T + e^{\mathrm{trans}}{}_T$, assuming for simplicity the underlying problem to be linear. This asks for

- error propagation in space (global pollution effect)

- interaction of physical error sources (local sensitivity analysis)

The effect of the cell residual ρ_T on the local error $e_{T'}$ at another cell T' is essentially governed by Green's function of the continuous problem. Capturing this dependence by *numerical* evaluation is the general philosophy underlying our approach to error control. In practice it is mostly impossible to determine the complex error interaction by analytical means, it rather has to be detected by computation. This results in a feed-back process in which error estimation and mesh adaptation go hand-in-hand leading to economical discretisation for computing the quantities of interest.

This procedure of employing an auxiliary (dual) problem to derive weighted *a posteriori* error estimates for finite element discretisations of (nonlinear) variational equations (VE) has been proposed by Johnson and his co-workers (see, e.g., Eriksson et al. [29]) and a refined strategy, the DWR-method, has been successfully used for various problems (see, e.g., Becker [7], Kanschat [42], Suttmeier [66] and Rannacher et al. [53, 5, 9] for a survey). In many applications this framework can be applied also to variational inequalities, by transforming the original formulation into a nonlinear equation. Examples are penalty methods for contact or obstacle problems (see, e.g., Carstensen et al. [20], Hansbo and Johnson [36]), or regularizations for inelastic materials (see, e.g., Rannacher and Suttmeier [57]). In this work, we demonstrate, how the techniques to derive residual based error estimators can be extended directly to the original variational inequalities by employing a suitable adaptation of Nitsche's duality argument. (cf. Natterer [51]).

Studying the duality techniques (Nitsche trick) used to obtain improved *a priori* error estimates in the L^2-norm, gives ideas how to derive weighted residual based *a posteriori* estimates for FE-discretisations of variational equalities (see, e.g., Rannacher et al. [53, 5, 9]). We employ the analoguous approach to get comparable results in the context of variational inequalities.

To this end we first sketch the *a priori* error analysis of a simple example, found in Natterer [51], where a duality argument is exploited to obtain improved L^2-error estimates in an inequality setting. Generalising the techniques therein allows for the construction of a framework for deriving *a posteriori* error bounds for FE-approximations of a class of variational inequalities.

As a basic model situation, we consider a one dimensional problem of Signorini type on $\Omega = (0,1) \subset \mathbb{R}^1$, which in classical notation reads

$$u - u'' = f \text{ in } \Omega \,,$$
$$u(0) \geq 0 \,,\; u(1) \geq 0 \,,\quad u'(0) \leq 0 \,,\; u'(1) \geq 0 \,,$$
$$u(0)u'(0) = u(1)u'(1) = 0 \,.$$

Introducing the bilinear forms

$$a(v,\varphi) := \int_\Omega (v\varphi + v'\varphi') \, dx \,, \quad (v,\varphi) := \int_\Omega v\varphi \, dx \,,$$

with corresponding norms $\|.\|_V$ and $\|.\|$, the weak formulation is

$$u \in K : \quad a(u, \varphi - u) \geq (f, \varphi - u) \qquad \forall \varphi \in K \,, \tag{1.1}$$

where $V = H^1(\Omega)$ and $K = \{v \in V | v(0) \geq 0 \,, v(1) \geq 0\}$.

With $0 = x_0 < x_1 < \cdots < x_n = 1$ and $h = \max(x_i - x_{i-1})$ we define the discrete space $V_h = \{v \in C[0,1] \mid v \text{ is linear on } [x_i, x_{i-1}]\}$. Setting $K_h =$

$K \cap V_h$ the discrete version of (1.1) is

$$u_h \in K_h : \quad a(u_h, \varphi - u_h) \geq (f, \varphi - u_h) \qquad \forall \varphi \in K_h . \tag{1.2}$$

A priori error analysis: The discussion starts with stating the result of an error estimate in the energy norm, which can be found, e.g., in Falk [30],

$$\|u - u_h\|_V \leq ch .$$

Using the definitions

$$B_h = \{x \in \{0, 1\} \mid u_h(x) = 0\} ,$$

$$G = \{v \in V | v \geq 0 \text{ on } B_h \text{ and } a(\mathcal{U} - u, v + u_h - u) \geq 0\} ,$$

one can use a duality argument

$$z \in G : \quad (u - u_h, \varphi - z) \leq a(\varphi - z, z) \qquad \forall \varphi \in G , \tag{1.3}$$

to obtain an improved L^2-estimate. $\mathcal{U}$ appearing in the definition of G is the solution of the unrestricted problem

$$\mathcal{U} \in V : \quad a(\mathcal{U}, \varphi) = (f, \varphi) \qquad \forall \varphi \in V .$$

The introduction of $\mathcal{U}$ offers the possibility of a geometric interpretation of the concept as illustrated in Natterer [51]. From practical point of view, the use of $\mathcal{U}$ is purely formal to have a compact notation. All terms involving $a(\mathcal{U}, .)$ can be alternatively expressed by $(f, .)$.

For the dual solution z of (1.3) one proves the estimates $\|z\| \leq \|u - u_h\|$ and $\|z''\| \leq 2\|u - u_h\|$, and consequently, $\|z - z_h\| \leq ch^2\|u - u_h\|$ and $\|z - z_h\|_V \leq ch\|u - u_h\|$, where z_h denotes the linear interpolant of z. Finally, one comes up with

$$(u - u_h, u - u_h) \leq a(u - u_h, z - z_h) + a(\mathcal{U} - u, z - z_h)$$
$$\leq ch\|u - u_h\|_V\|u - u_h\| + \|u - \Delta u - f\|h^2\|z''\| ,$$

yielding the desired estimate $\|u - u_h\| \leq ch^2$.

In the following, we employ the ideas sketched above to derive *a posteriori* error estimates for problems given in an abstract setting

$$A(U, \varphi - U) \geq F(\varphi - U) \qquad \forall \varphi \in \mathbf{K} , \tag{1.4}$$

where $A(.,.)$ and $F(\cdot)$ are a continuous bilinear and linear form, respectively, on a closed, convex subset $\mathbf{K} \subset \mathbf{V}$ of the Hilbert space $\mathbf{V}$. In our applications from structural mechanics, as indicated below, we assume that $\mathbf{V}$ is a closed subspace of $(H^1(\Omega))^N$ and A is uniformly elliptic on $\mathbf{V}$. Here, and in what follows, $W^{m,p} = W^{m,p}(\Omega)$ denotes the standard Sobolev space of L^p-functions with derivatives in $L^p(\Omega)$ up to the order m, where Ω is a bounded domain in $\mathbb{R}^d$, $d = 1, 2, 3$. The subscript in $W_0^{m,p}$ indicates zero boundary conditions and we write $H^m = W^{m,p}$ for $p = 2$. Furthermore, $(.,.)$ represents the L^2 inner product and $\|.\|$ the corresponding norm.

We apply the finite element method on decompositions $\mathbb{T}_h = \{T_i \mid 1 \leq i \leq N_h\}$ of Ω consisting of N_h triangular, quadrilateral or hexahedral elements T_i, satisfying the usual condition of shape regularity. For ease of mesh refinement and coarsening, hanging nodes are allowed in our implementation. The width of the mesh $\mathbb{T}_h$ is characterised in terms of a piecewise constant mesh size function $h = h(x) > 0$, where $h_T := h_{|T} = \text{diam}(T)$.

Using this notation, the solution $U \in \mathbf{K}$ is approximated by $U_h \in \mathbf{K}_h$ with

$$A(U_h, \varphi - U_h) \geq F(\varphi - U_h) \qquad \forall \varphi \in \mathbf{K}_h, \tag{1.5}$$

where $\mathbf{V}_h$ is a finite element space on $\mathbb{T}_h$ and $\mathbf{K}_h \subset \mathbf{V}_h$ an appropriate subset. These discrete spaces approximate $\mathbf{V}$ and $\mathbf{K}$ respectively. If not stated explicitly $\mathbf{V}_h$ is constructed by the standard d-linear shape functions.

We propose a framework for deriving a *posteriori* error estimates for the scheme (1.5), where the error $e = U - U_h$ is measured in terms of a linear functional $J(\cdot)$ defined on $\mathbf{V}$ or a suitable subspace. The main ingredients are a generalisation of the Galerkin orthogonality relation in the context of variational equations and a suitable duality argument. The structure of our estimates is

$$|J(e)| \leq F(Z - Z_h) - A(U_h, Z - Z_h),$$

where Z is the solution of an auxiliary problem similar to (1.3) and Z_h denotes a suitable corresponding approximation. This can be split into

$$|J(e)| \leq \sum_{T \in \mathbb{T}_h} \omega_T \rho_T(F, U_h) =: \eta(U_h),$$

with the local *residuals* ρ_T depending on the data of the underlying problem
and the computed solution U_h. We emphasize, that the local *weights* $\omega_T = \omega_T(Z(J))$ depend on the measure $J(\cdot)$. In more traditional approaches, one
only chooses $\omega_T = $ const.

The effect of using $\omega_T(Z(J))$ is explained by Figure 1.5. Here, an adaptively
refined grid for an obstacle problem on a slit domain is shown. The *weighted*
approach balances the local mesh size around the contact zone, the corner
singularity and the quantity of interest, which in this example is a point value
in the lower left quarter of the domain. Compared to conventional strategies
(ZZ), the distribution of degrees of freedom yields a more accurate approximation.

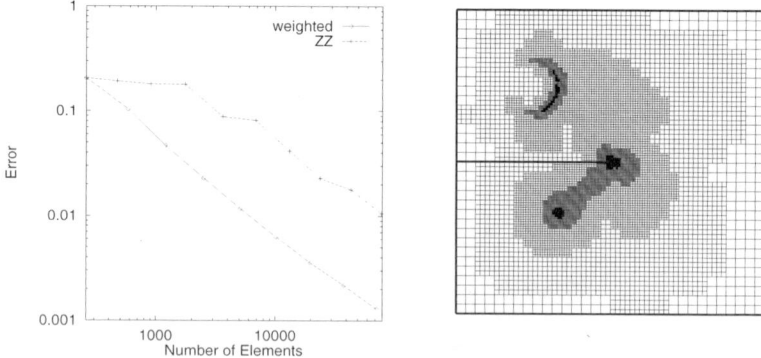

Figure 1.5: Relative error for the obstacle problem (left). Structure of grids
produced using the *weighted* approach (right).

All numerical tests we performed, show that our estimators yield an asymptotically correct description of the error behaviour. Furthermore we remark
that the overestimation-factors are about $2 - 4$. In addition, comparison with
conventional strategies shows the high degree of efficiency of our estimators.

The numerical results presented throughout this work are obtained by FE-
implementations based on the DEAL-library [21].

Due to our new object oriented approach we are able to treat relevant problems
of computational mechanics on machines with the restrictions of a workstation.
One advantage of our concept is the *flexibility*. For example, in the 2D-case

the discretisation of linear elasticity problems the memory requirements are moderate, since we have to treat only the two unknowns for the displacement. So we can use a traditional matrix to represent the linear operator describing the discrete problem. In contrast in the 3D-case the memory requirements are much higher. In this situation, we only store information about the discretisation on the lowest level. The required element matrices on each cell are simply computed by using the hierarchical structure of our meshes. We want to point out, that the choice of the representation of the operator only requires local implementation. The global code has not to be changed. This is a very essential point to construct reliable code.

The solutions on very fine (adaptive) meshes with about 200,000-500,000 cells are taken as *reference* solutions U_{ref} for determining the relative errors

$$E^{\mathrm{rel}} := |J(U_h) - J(U_{\mathrm{ref}})|/|J(U_{\mathrm{ref}})|\,,$$

on coarser meshes, while

$$Ratio := \frac{\eta(U_h)}{|J(U_{\mathrm{ref}}) - J(U_h)|}\,,$$

are the overestimation factors of the error estimators η.

Let an error tolerance TOL or a maximal number of cells N_{max} be given. Starting from some initial coarse mesh the refinement criteria are chosen in terms of the *local error indicators* $\eta_T := \omega_T \rho_T$. Then for the mesh refinement, we use in the majority of cases the following *fixed fraction* strategy: In each refinement cycle, the elements are ordered according to the size of η_T and then a fixed portion of about 30% of the elements with largest η_T is refined resulting in approximately a doubling of the number N of cells. This process is repeated until the stopping criterion $\eta(u_h) \leq TOL$ is fulfilled or N_{max} is exceeded. A more detailed discussion about mesh refinement strategies is given below and can be found, e.g., in Rannacher [53].

In the numerical tests, we compare the *weighted* estimator against the traditional approach of Zienkiewicz and Zhu [74]. This error indicator for finite element models in structural mechanics is based on the idea of higher–order recovery of the stresses σ by local averaging. The element–wise error $\|\sigma - \sigma_h\|_T$ is thought to be well represented by the auxiliary quantity $\eta_T := \|\mathcal{M}_h \sigma_h - \sigma_h\|_T\,,$

where $\mathcal{M}_h \sigma_h$ is a local (super-convergent) approximation of σ. The corresponding (heuristic) global error estimator reads

$$\|\sigma - \sigma_h\| \approx \eta_{ZZ} := \left(\sum_{T \in \mathbb{T}_h} \|\mathcal{M}_h \sigma_h - \sigma_h\|_T^2 \right)^{1/2}, \qquad (1.6)$$

where σ and σ_h are expressed in terms of derivatives of U and U_h, respectively.

Overview: Parts of the introduction were taken from Rannacher & Suttmeier [58] and Suttmeier [68]. The further organisation of this work is as follows.

First, material taken from Suttmeier [66], we briefly review the physical models we use for elasto-plasticity throughout this book.

In Chapter 3, material partly taken from Blum & Suttmeier [13] and Suttmeier [66], we recall the concept of the *dual-weighted-residual method* at a problem in plasticity theory, which leads to a variational inequality, but can be rewritten as an equality by means of a nonlinear projection onto the yield surface.

In Chapter 4, material partly taken from Suttmeier [67], mixed finite element schemes for solving problems in continuum mechanics are discussed, which are often used to obtain higher-order approximation for the stresses. However, it is viewed as a drawback of these methods that one has to construct suitable non-standard finite element spaces to obtain stable discretisations. An alternative approach is to augment the original mixed formulation by least-squares-like terms, allowing the use of standard finite element spaces for the approximation. For this scheme weighted a posteriori error estimates can be derived by duality arguments. The underlying dual problems have to be chosen according to the mesh-dependent stabilisation parameters in order to guarantee optimal-order estimation of the least-squares terms. The construction of a stabilised finite element scheme and the appropriate adaptive algorithm is developed for a model example in elasticity theory.

In Chapter 5, material partly taken from Blum & Suttmeier [13] and Suttmeier [68], we demonstrate at the so-called obstacle problem, how a problem formulated in form of a variational inequality can be attacked directly to derive *a posteriori* error estimates analogously to the example in the introduction. Our procedure is performed in three steps, following the approach in the context of variational equalities, i.e., a priori error analysis is first done for the energy norm, then extended to L^2-norm estimates finally leading to *a posteriori* error bounds.

In Chapter 6, material partly taken from Blum & Suttmeier [12] and Suttmeier

[68], we consider Signorini's problem, a fundamental model situation for contact problems in elasticity, which is the basis for applications in the field of highspeed machining presented below. This problem is to be solved by the finite element Galerkin method on adaptively optimized meshes.

In Chapter 7, material partly taken from Suttmeier [69, 68], the example from elasto-plasticity theory is revisited. In this case we have to deal with a system of unknowns yielding a mixed variational setting. Furthermore in contrast to the previous examples, the imposed inequality constraints are nonlinear.

In Chapter 8, material partly taken from Suttmeier [69, 68], the experiences from the three examples treated above are collected. We propose a framework for deriving *a posteriori* error estimates for the abstract scheme (1.5).

In Chapter 9, we will discuss a further application of the theory of VI's, namely the torsion problem. It turns out to be convenient to treat this problem by a Lagrangian formalism. The approach offers several alternatives for the numerical analysis of variational inequalities. We mention the iterative solution process of the discrete problems and focus on new possibilities for *a priori* error analysis.

In Chapter 10, material partly taken from Suttmeier [68], motivated by the results of the previous chapter, we come back to the obstacle problem, which we treat here by the Lagrangian formalism and derive *a posteriori* error estimates for the corresponding FE-discretisation employing the general framework from Chapter 8.

In Chapter 11, we extend our studies on finite element Galerkin schemes for elliptic variational inequalities of first to the one of second kind. Especially we perform the corresponding *a posteriori* error analysis for a simple friction problem and a model flow of a Bingham fluid.

In Chapter 12, we indicate how the abstract framework presented above for stationary problems may be appplied to non-stationary processes. This is illustrated at a parabolic model situation, which may be interpreted as a time-dependent obstacle problem.

In Chapter 13, material partly taken from Rannacher & Suttmeier [58] and Suttmeier [68], we present the application of our strategies to processes in the field of highspeed machining. This is motivated by our membership in the research group *ForscherGruppe FreiFormFlächen* supported by the Deutsche Forschungsgemeinschaft. In collaboration with scientists from mechanical engineering, we provide the numerical analysis for grinding and milling processes.

Furthermore we show some results concerning adaptive mesh design for FE models in elasto-plasticity to demonstrate our techniques to work within a time-stepping scheme on locally varying meshes in space including hanging nodes.

In Chapter 14, material partly taken from Blum et al. [11], we present ideas and recent results for a cascadic multigrid algorithm for variational inequalities. Eventually, Chapter 15 collects some conclusions.

An appendix, containing several practical aspects, concludes this book.

Chapter 2

Models in elasto-plasticity

In this chapter, we briefly review the physical models we use for elasto-plasticity throughout this book, since the treatment of the related problems requires *Finite Elements made to measure* from several point of views.

From mathematical point of view elasto-plastic problems are optimisation problems, where the stress field is subjected to inequality constraints which have to be fulfilled in the pointwise sense. The construction of a numerical algorithm based on this sight seems to be involved, as discussed in Suttmeier [66]. A more straight-forward approach for a corresponding FE-scheme starts from the *primal-mixed* formulation, discussed in Chapter 3. Here, the problem can be rewritten as an equality by means of a nonlinear projection onto the yield surface.

On the other hand, even very simple examples in the elasto-plastic case show, that discontinuities of the displacement can occur. In contrast, in [10] there is proven, that the stresses belong to H^1_{loc}. Due to these regularity results it seems to be more adequate to consider the *dual-mixed* approach, discussed in Chapter 4, since derivatives of u are not entering this formulation.

And last, but not least, in general the available a priori results are not sufficient to get useful information about the accuracy of a numerical approximation. Especially in the case of perfect plasticity the known a priori estimates for the discretisation error given in the energy norm are very weak, and in large parts of the domain they are asymptotically wrong. We mention the results of Falk & Mercier [31] and Johnson [39, 38], where an $\mathcal{O}(\sqrt{h})$ convergence is proven for the L^2-norm of the stresses. Under the assumption of smooth data our numerical experience is that the $\mathcal{O}(h)$ convergence dominates. Weaker convergence appears only in the transition zones between elastic and plastic behaviour. Therefore we see that it is urgent to develop efficient adaptive algorithms to catch the regions responsible for the poor convergence.

2.1 Governing equations

The behaviour of elastic and inelastic materials subjected to external forces can be described by a system of partial differential equations. This system consists of two components.

Equilibrium law of forces

We denote by $\Omega \subset \mathbb{R}^d (d = 2$ or $3)$ a bounded domain occupied by an arbitrary solid. Along a part Γ_u of the boundary $\partial\Omega$ the material is fixed. Further the body is subjected to a body force with density f and a surface traction g along $\Gamma_\sigma = \partial\Omega \setminus \Gamma_u$.

Due to the external forces one observes a translation u of a point x in the unloaded case to the place ξ in the deformed state. So there is the relation $\xi_i = x_i + u_i(x,t)$, where u_i are the components of the displacement. The deformation of the body can be described by the deformation tensor

$$\tilde{\varepsilon}(u) := \frac{1}{2} \left(\nabla u + \nabla u^T + \nabla u^T \nabla u \right).$$

Supposing only small deformations $|\nabla u|$ we neglect the quadratic term $\nabla u^T \nabla u$ and regard $\varepsilon(u)$ as the *linearised deformation tensor*

$$\varepsilon(u) := \frac{1}{2} \left(\nabla u + \nabla u^T \right).$$

There arise inner forces due to the deformation, which is caused by the external loads. These forces are responsible for the new equilibrium state and can be described by the symmetric *stress tensor* σ.

Taking into account the conservation of mass, momentum and energy one derives the equation

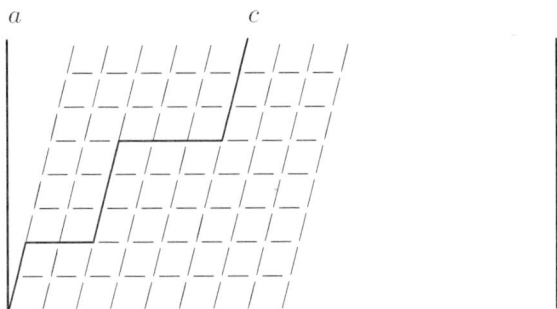

Figure 2.1: Schematic representation: gliding movement of the atoms causes plastic behaviour

$$\frac{\partial^2 u}{\partial t^2} - \operatorname{div} \sigma - f = 0.$$

Under the assumption that acceleration effects can be neglected we state the static equilibrium law

$$-\operatorname{div} \sigma = f.$$

This relation gives us three equations to determine the six components of the stress tensor. The deformation a body undergoes due to external load depends on the properties of the solid. So the equations still missing are given by a material law, in form of a relation between ε and σ.

Material law

The micro structure of a solid can be regarded as a grid of interconnected atoms. If there is a deformed grid with an unchanged inner structure – sketched in Figure 2.1, we are talking of *elastic* behaviour. The body goes back into its original state if the external load vanishes. If there is a deformed grid, where the inner structure changes, we are talking of *plastic* behaviour. These inner changes occur due to a gliding movement of the atoms. This can be explained with the help of Figure 2.1. In the unloaded case we look at atoms which are all on the straight line a. After plastic behaviour has taken place we find the atoms on a distorted line c. During the gliding process mechanical energy is lost and the temperature changes locally. One observes a

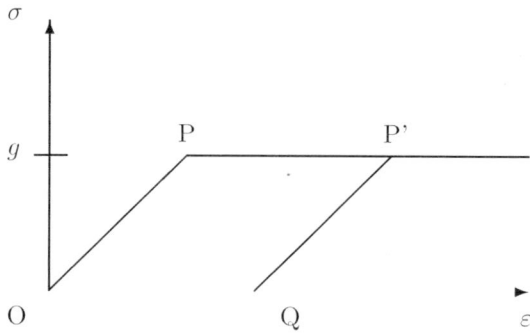

Figure 2.2: Stress-strain diagram

dissipation of energy. If now the external load vanishes, one obtains a body with a changed geometry.

The difference between elastic and plastic material behaviour is essentially described by certain norms of the inner stress. As long as these norms are small elastic behaviour dominates. If the values are above a certain limit, one observes plastic behaviour. To get a mathematical model we investigate in the following the relation between elastic and perfect plastic material behaviour. This can be done with the help of Figure 2.2, where the material law of an one dimensional rod is sketched.

First we consider the case, where the deformation to the state P is small enough. $|\sigma|$ remains below the limit g or written in terms of the so called flow function $\mathcal{F}$ we have $\mathcal{F}(\sigma) = |\sigma| - g < 0$. Removing the external load the origin O is reached again. Now let $1/A$ be the slope of the line OP. Then one can write the elastic material relation in the form $\varepsilon = A\sigma$. Generalising this concept for three dimensions, we arrive at $\varepsilon_{ij}(u) = A_{ijkh}\,\sigma_{kh}$. In the case, where the properties of a solid can be regarded as isotropic, one can write the material law – law of Lamé and Navier – in the form

$$\sigma = 2\mu\varepsilon + \lambda\,\mathrm{tr}(\varepsilon),$$

where λ and μ are certain material dependent constants.

If the deformation ε increases further after arriving at state P, the solid is unable to produce inner stresses larger than g. Therefore the gliding movement takes us to the state P'. After vanishing of the external forces the solid obeys the linear elastic law. It chooses a parallel line to OP starting in P' to reach the new unloaded state Q. In other words we now have in the unloaded state a body with different geometry compared to the initial state. To investigate the relation between σ and ε we consider infinitesimal small changes of the corresponding quantities:

$$d\varepsilon = A\,d\sigma + \tilde{\lambda}\,, \text{ with } \tilde{\lambda} = 0 \text{ if } \sigma < g\,,$$
$$\tilde{\lambda} = 0 \text{ if } d\sigma < 0\,, \sigma = g\,,$$
$$\tilde{\lambda} \geq 0 \text{ if } \sigma = g\,, d\sigma = 0\,.$$

Assuming these changes to happen during the time Δt and considering the limit process $\Delta t \to 0$, one obtains

$$\dot{\varepsilon} = A\dot{\sigma} + \lambda\,;\, \lambda = 0 \text{ if } \sigma < g\,,$$
$$\lambda = 0 \text{ if } \dot{\sigma} < 0\,, \sigma = g\,,$$
$$\lambda \geq 0 \text{ if } \sigma = g\,, \dot{\sigma} = 0.$$

These (in-)equality constraints can equivalently be written in the form

$$\lambda\,(\tau - \sigma) \leq 0 \quad \forall\, \tau \leq g\,,$$
$$\lambda\,\dot{\sigma} = 0\,.$$

Generalising this mathematical model for the three dimensional case we choose a function $\mathcal{F}$, continuous and convex in σ_{ij}, to describe the flow rule and so the material law reads

$$\dot{\varepsilon}_{ij}(u) = A_{ijkh}\,\dot{\sigma}_{kh} + \lambda_{ij}\,,$$
$$\lambda_{ij}\,(\tau_{ij} - \sigma_{ij}) \leq 0 \quad \forall\, \tau \text{ with } \mathcal{F}(\tau_{ij}) \leq 0\,,$$
$$\lambda_{ij}\,\dot{\sigma}_{ij} = 0\,.$$

Flow rule

The inner stress can be divided into two parts. The *normal stress* is caused by pure volume changes and due to shape changes there arises *shear stress*. Regarding σ_{ij} as a linear mapping we denote by σ_i its eigen values. With the

help of these quantities one can describe the maximum values τ_i of the shear
stresses with

$$\tau_1 = \frac{1}{2}(\sigma_2 - \sigma_3), \ \tau_2 = \frac{1}{2}(\sigma_3 - \sigma_1), \ \tau_3 = \frac{1}{2}(\sigma_1 - \sigma_2).$$

A general flow condition was stated by Tresca in 1868:

> Plastic deformation takes place, if the maximum value of the shear
> stresses reaches a certain (dependent on material and temperature)
> value.

Taking into account our model of perfect plasticity, we have flow behaviour if
one of the following equations is fulfilled:

$$|\tau_1| = \frac{1}{2}|\sigma_2 - \sigma_3| = g \,,$$

$$|\tau_2| = \frac{1}{2}|\sigma_3 - \sigma_1| = g \,,$$

$$|\tau_3| = \frac{1}{2}|\sigma_1 - \sigma_2| = g \,.$$

In this case the domain $\{\sigma_{ij}, \ \mathcal{F}(\sigma_{ij}) \leq 0\}$ is simply given by planes. Von
Mises approximated this polyhedron by a cylinder with elliptic cross section.

$$\frac{1}{2}\left[(\sigma_1 - \sigma_2)^2 + (\sigma_2 - \sigma_3)^2 + (\sigma_3 - \sigma_1)^2\right] - g^2 = 0. \tag{2.1}$$

To get a more compact formula we remark that a given stress state can be
divided into two parts. First there is the *isotropic* stress state in the form

$$\sigma_{\mathrm{iso}} = \begin{pmatrix} \sigma_m & 0 & 0 \\ 0 & \sigma_m & 0 \\ 0 & 0 & \sigma_m \end{pmatrix} \quad \text{with } \sigma_m = \frac{1}{3}\operatorname{tr}\sigma.$$

It is related to the stress state in a fluid free of inner friction. The *deviatoric*
stress state σ^D is given by

$$\sigma_{ij}^D = \sigma_{ij} - \sigma_m\delta_{ij}.$$

Considering (2.1) one can equivalently write

$$|\sigma^D| = g.$$

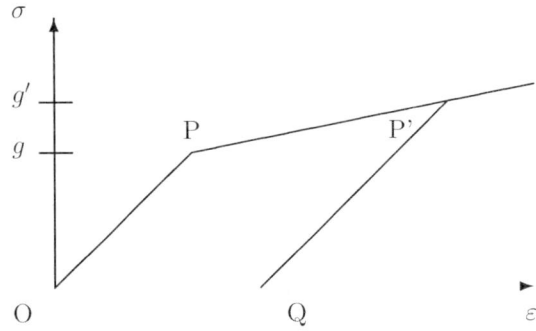

Figure 2.3: Stress-strain diagram, linear hardening

So the deviator plays an important role in the description of plastic behaviour with the von Mises flow function, since this flow rules are given in terms of σ^D. In the context of perfect plasticity we have $\mathcal{F}(\sigma) = |\sigma^D| - g \leq 0$.

Hardening

As mentioned above apart from variations in the geometry, other properties of a solid change if there is plastic deformation. We introduce the term *hardening* to denote the change in the flow behaviour of a body during a loading process. We use figure (2.3) to explain such a property. Like in the perfect plastic case we arrive after a plastic loading at the state Q, which describes the *new* body. If this new one is now subjected to external forces, it is now linear elastic up to a higher flow value g' compared to g of the old state O.

Under certain conditions the anisotropy which arises due to plastic deformation can be captured by combining isotropic and kinematic hardening. The flow function describing such a case reads

$$\mathcal{F}(\tau, \eta) = |\tau^D - \gamma_{kin}\eta_{kin}| - (g + \gamma_{iso}\eta_{iso}),$$

where $\gamma_{kin} > 0$ and $\gamma_{iso} > 0$ are given constants. Introducing the plastic strain $\dot{\varepsilon}_p = \dot{\varepsilon} - A\dot{\sigma}$, the hardening rule for isotropic (strain-) hardening used in mechanics is $\dot{\eta}_{iso} = |\dot{\varepsilon}_p|$.

The so called kinematic hardening rule is given by $\dot{\eta}_{kin} = \gamma_{kin}\dot{\varepsilon}_p$. This behaviour corresponds to a translation of the midpoint of the flow domain.

2.2   Examples

In this section, we simply adapted two examples from Suquet [65] in one dimension to the Hencky-type problem. We see, that the solutions for u can have discontinuities or are not uniquely defined.

Discontinuity of the displacements can occur

Suppose on $\Omega = (-1; 1)$ the elastic relation $\sigma = u'$, flow rule $|\sigma| \leq 1$, boundary conditions $u(-1) = 0$, $u(1) = 0$ and equilibrium law $-\operatorname{div} \sigma = -\sigma' = f$, where with $3 < \kappa < 4$ we choose

$$
f(x) = \begin{cases} -\kappa(1-x), & x \geq 0, \\ \kappa(1+x), & x < 0. \end{cases}
$$

Because $-\sigma' = f$ the static admissible stress fields τ are

$$
\tau(x) = C_\tau + \begin{cases} \kappa(x - \frac{x^2}{2}), & x \geq 0, \\ -\kappa(x + \frac{x^2}{2}), & x < 0. \end{cases}
$$

If there holds $-1 \leq C_\tau \leq 1 - \kappa/2$, then $|\tau(x)| \leq 1$ is fulfilled. The unique solution is

$$
\sigma(x,t) = -1 + \begin{cases} \kappa(x - \frac{x^2}{2}), & x \geq 0, \\ -\kappa(x + \frac{x^2}{2}), & x < 0. \end{cases} \tag{2.2}
$$

To see this, one has to prove that for σ from (2.2) $(\sigma, \tau - \sigma) \geq 0$ is true:

$$
(\sigma, \tau - \sigma) = 2 \int_0^1 (-1 + \kappa(x - \frac{x^2}{2}))(C_\tau + 1) \, dx = (\frac{2\kappa}{3} - 2)(C_\tau + 1).
$$

So if there holds $-1 \leq C_\tau \leq 1 - \kappa/2$, we observe

$$
(\sigma, \tau - \sigma) = (\frac{2\kappa}{3} - 2)(C_\tau + 1) \geq 0,
$$

and therefore the function given in (2.2) is the solution. Furthermore we have $\frac{du}{dx} = \sigma$, and from this we conclude

$$
u(x) = \begin{cases} C_+ + \kappa(\frac{x^2}{2} - \frac{x^3}{6}), & x > 0, \\ C_- - \kappa(\frac{x^2}{2} + \frac{x^3}{6}), & x < 0. \end{cases}
$$

Because of the boundary conditions we have $C_+(t) = -\frac{\kappa}{3}$ and $C_-(t) = +\frac{\kappa}{3}$.
Summarising, the (unique) solution of the given problem is:

$$\sigma(x) = -1 + \begin{cases} \kappa(x - \frac{1}{2}x^2), & x > 0, \\ -\kappa(x + \frac{1}{2}x^2), & x < 0, \end{cases}$$

$$u(x) = \begin{cases} -\frac{1}{3} + \kappa(\frac{1}{2}x^2 - \frac{1}{6}x^3), & x > 0, \\ \frac{1}{3} - \kappa(\frac{1}{2}x^2 + \frac{1}{6}x^3), & x < 0. \end{cases}$$

We observe a discontinuity in the displacement at point $x = 0$ (cf. Figure 2.4).

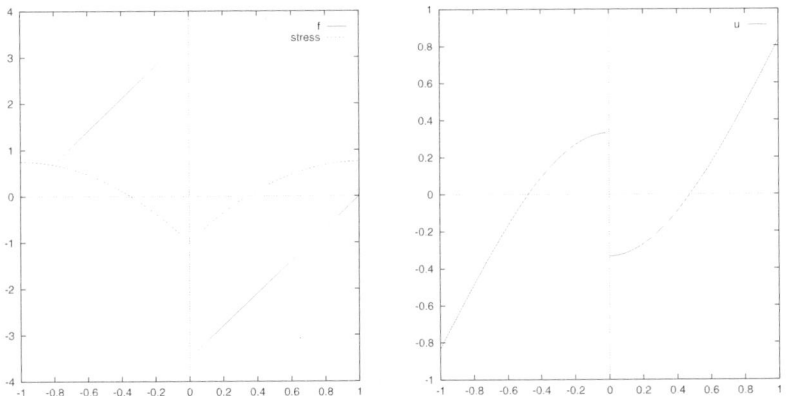

Figure 2.4: Load f, stress σ (left) and displacement u (right) of the first example.

Non-uniqueness of the displacements can occur

Suppose on $\Omega = (0, 1)$ the elastic relation $\sigma = u'$, flow rule $|\sigma| \le 1$, equilibrium law $\operatorname{div}\sigma = \sigma' = 0$ and boundary conditions $u(0) = 0$, $u(1) = \lambda > 0$. The material law can be written

$$(\sigma, \tau - \sigma) - (u', \tau - \sigma) \ge 0$$
$$(\sigma, \tau - \sigma) + (u, \tau' - \sigma') - \lambda(\tau(1) - \sigma(1)) \ge 0.$$

Because of $\sigma' = 0$ there holds $\sigma = s \in \mathbb{R}$. Let $P = \{s \in \mathbb{R}, |s| \le 1\}$. So s is determined by

$$(s, t - s) - \lambda(t - s) \ge 0 \qquad \forall t \in P$$

or equivalently by

$$s = \min \left\{ \frac{1}{2}t^2 - \lambda t, \ t \in P \right\}.$$

We conclude that for $\lambda > 1$ the plastic solution is $\sigma = 1$. A corresponding displacement is described by

$$(1, \tau - 1) + (u, \tau') \geq [\lambda \tau(1) - \lambda]$$

$$(1, \tau - 1) - \int u' \tau \geq -\lambda.$$

Choosing $\lambda > 2$ and arbitrary $x_0 \in (0, 1)$, one observes that the functions

$$u = \begin{cases} 0 & \text{if } x < x_0 \\ \lambda & \text{if } x \geq x_0 \end{cases}$$

fulfil $-2 - \int u' \tau \geq -\lambda$, and because of $(1, \tau - 1) \geq -2$ each of the above functions is a solution to the plastic problem.

Chapter 3

The dual-weighted-residual method

The procedure of employing an auxiliary
(dual) problem to derive weighted *a posteriori* error estimates for FE-discretisations of
(nonlinear) variational equations has become
standard during the last years. In this chapter, we recall this concept in more detail at a
problem in plasticity theory, which leads to
a variational inequality, but can be rewritten as an equality by means of a nonlinear
projection onto the yield surface. In contrast
to this example for most relevant problems
in the theory of variational inequalities no
conversion to nonlinear equalities, suitable
to construct efficient numerical schemes, is
known. We shall demonstrate in the subse-
quent chapters, that this technique to derive
weighted *a posteriori* estimators can nevertheless be carried over to the case of inequalities by carefully adapting the duality argument.

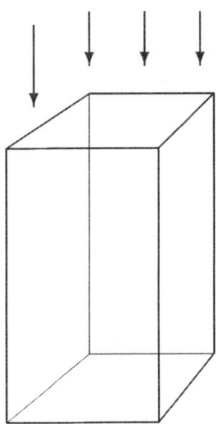

Figure 3.1: Geometry sketch
of the Strang example

3.1   A model situation in plasticity

We study a model problem posed by Strang [64]. The physical problem is that of an infinitely long straight pipe with quadratic cross–section $\Omega \subset \mathbb{R}^2$, filled with plastic material with linear hardening adherent to the walls and subjected to a volume force f acting in vertical direction (see Figure 3.1). The mathematical model seeks for a scalar displacement u in the vertical direction and a stress vector $\sigma = (\sigma_1, \sigma_2)$ as functions on Ω. The plastic behaviour of the material is taken into account by the nonlinear restriction $|\sigma| \leq 1 + \alpha\xi$, with $\alpha > 0$ and $\xi = \xi(\nabla u)$. This results in the system

$$-\operatorname{div}\sigma = f, \quad \sigma = \Pi_\xi \nabla u \quad \text{in } \Omega, \tag{3.1}$$
$$u = 0 \quad \text{on } \partial\Omega,$$

where Π_ξ denotes the pointwise projection onto the circle with radius $1 + \alpha\xi$. In order to give a weak form for (3.1), we set

$$L^2(\Omega)^2 := L^2(\Omega, \mathbb{R}^2),$$
$$\Pi H := \left\{ (\tau, \eta) \in L^2(\Omega)^2 \times L^2(\Omega), |\tau| - (1 + \alpha\eta) \leq 0 \right\},$$
$$V := \left\{ u \in H^1(\Omega), \ u = 0 \text{ on } \partial\Omega \right\}.$$

Now, similar to the approach in Johnson [40], the solution $\{(\sigma, \xi), u\} \in \Pi H \times V$ is determined by the variational inequality

$$(\sigma, \tau - \sigma) + \alpha(\xi, \eta - \xi) - (\nabla u, \tau - \sigma) + (\sigma, \nabla\varphi) \geq (f, \varphi), \tag{3.2}$$

for any pair $\{(\tau, \eta), \varphi\} \in \Pi H \times V$. Existence and uniqueness of the solution have been proven, e.g., by Johnson [40].

Choosing $\eta = \xi$ and $\varphi = 0$, we see that σ is the projection of ∇u onto the circle with radius $1 + \alpha\xi$. Additionally employing the relation $\xi = |\nabla u - \sigma|$ (cf. [40]) one obtains

$$\sigma = C(\nabla u), \quad \text{a.e. in } \Omega, \tag{3.3}$$

with the function

$$C(\tau) := \Pi_\xi \tau = \begin{cases} \tau, & \text{if } |\tau| \leq 1, \\ (1 - \gamma)\dfrac{\tau}{|\tau|} + \gamma\tau, & \text{if } |\tau| > 1, \end{cases}$$

where $0 < \gamma < 1$ is obtained by eliminating the parameter α from (3.2). Eventually, the variational inequality (3.2) can be transformed into a nonlinear variational equation

$$(C(\nabla u), \nabla \varphi) = (f, \varphi) \qquad \forall \varphi \in V. \tag{3.4}$$

Using the notation $V_h \subset V = H_0^1(\Omega)$ for the corresponding finite element subspaces, the approximate solution $u_h \in V_h$ is determined by the discrete equation

$$(C(\nabla u_h), \nabla \varphi) = (f, \varphi) \quad \forall \varphi \in V_h. \tag{3.5}$$

3.2 A posteriori error estimate

In order to estimate the error in the scheme (3.5) for general error measures given in terms of linear functionals $J(\cdot)$ defined on the space V, or on a suitable subspace containing V_h and the exact solution u, we proceed as follows (compare, e.g., Becker and Rannacher [8]).

Combining (3.4) and (3.5), we obtain the nonlinear Galerkin orthogonality relation

$$(C(\nabla u) - C(\nabla u_h), \nabla \varphi)$$
$$= \int_0^1 (C'(\nabla(su + (1-s)u_h))\nabla(u - u_h), \nabla \varphi) \, ds = 0, \quad (3.6)$$

for $\varphi \in V_h$, with the Jacobian $C'(\cdot)$ of the function $C(\cdot)$.

Now, suppose that the quantity $J(u)$ has to be computed. For representing the error $J(e) = J(u) - J(u_h)$, following a duality argument, known as "Aubin-Nitsche trick" from *a priori* analysis, we use the solution z of the linear dual problem

$$L(u, u_h; \varphi, z) = J(\varphi) \quad \forall \varphi \in V, \tag{3.7}$$

with the bilinear form

$$L(u, u_h; \varphi, \psi) := \int_0^1 (C'(\nabla(su + (1-s)u_h))\nabla \varphi, \nabla \psi) \, ds.$$

We assume that this *dual solution* is well defined. By the orthogonality relation (3.6), there holds

$$J(e) = L(u, u_h; e, z - z_h),$$

with a suitable approximation $z_h \in V_h$ of z. With standard techniques this can be exploited as follows. Elementwise integration by parts yields

$$J(e) = \sum_{T \in \mathbb{T}_h} \left\{ (f + \operatorname{div} C(\nabla u_h), z - z_h)_T - \tfrac{1}{2}(n \cdot [C(\nabla u_h)], z - z_h)_{\partial T} \right\},$$

where $[C(\nabla u_h)]$ denotes the jump of $C(\nabla u_h)$ across the interelement boundary. From this, we can conclude an error bound of the form

$$|J(e)| \leq \sum_{T \in \mathbb{T}_h} (\omega_T^1 \rho_T^1 + \omega_T^2 \rho_T^2), \tag{3.8}$$

with the local residuals and weights defined by

$$\rho_T^1 := h_T \| f + \operatorname{div} C(\nabla u_h) \|_T, \qquad \rho_T^2 := \tfrac{1}{2} h_T^{1/2} \| n \cdot [C(\nabla u_h)] \|_{\partial T},$$
$$\omega_T^1 := h_T^{-1} \| z - z_h \|_T, \qquad \omega_T^2 := h_T^{-1/2} \| z - z_h \|_{\partial T}.$$

3.3   Evaluation of a posteriori error bounds

In general, the weights ω_T^j, $j = 1, 2$, cannot be determined analytically, but have to be computed by solving the dual problem numerically on the available mesh. In practice, this is done as follows:

We replace the unknown exact solution u in the bilinear form $L(u, u_h; \cdot, \cdot)$ by the currently computed approximation u_h, and solve the corresponding perturbed dual problem by the same method as used in computing u_h, yielding an approximation $\tilde{z}_h \in V_h$ to the exact dual solution z,

$$L(u_h, u_h; \varphi, \tilde{z}_h) = J(\varphi) \quad \forall \varphi \in V_h. \tag{3.9}$$

The experiences in the case of the stationary Navier-Stokes equations (see Becker and Rannacher [8]) and for nonlinear elasto-plastic material behaviour (see Rannacher and Suttmeier [56]) indicate, that the perturbation of the dual problem is not critical in stable situations.

Next, one uses the interpolation estimates to control w_T^j, $j = 1, 2$, by

$$w_T^j \le w_T := C_{i,T} h_T \|\nabla^2 z\|_T \,, \tag{3.10}$$

for $z \in H^2(T)$. Then one evaluates the right hand side by simply taking second order difference quotients of the approximate dual solution $\tilde{z}_h \in V_h$,

$$w_T \approx \tilde{w}_T := \tilde{C}_{i,T} h_T^2 |\nabla_h^2 \tilde{z}_h(x_T)| \,, \tag{3.11}$$

where x_T is the midpoint of element T.

Eventually, we obtain the approximate *a posteriori* error bound

$$|J(e)| \approx \eta_{weight} := \sum_{T \in \mathbb{T}_h} \tilde{w}_T (\rho_T^1 + \rho_T^2) \,. \tag{3.12}$$

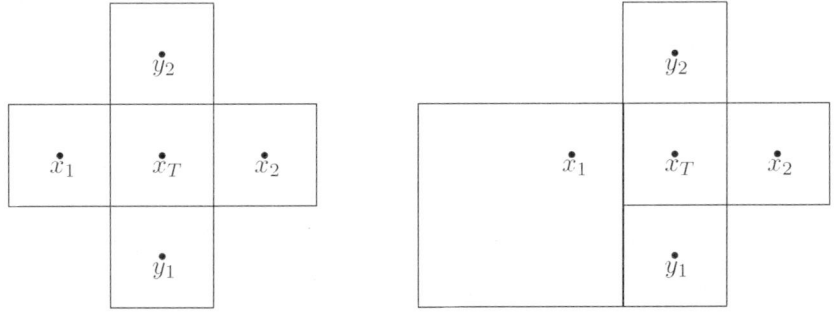

Figure 3.2: Sketch for illustrating the evaluation of $|\nabla_h^2 \tilde{z}_h(x_T)|$.

Remark: In our test calculations the second order derivatives of z are approximated as follows.

Let x_1, x_2, y_1, y_2 denote the midpoints of the neighbouring cells of T, see Figure 3.2 (left), then we evaluate

$$|\partial_x^2 z| \approx \frac{1}{2} \sum_{i=1,2} |\partial_x \tilde{z}_h(x_T) - \partial_x \tilde{z}_h(x_i)| \, |x_T - x_i|^{-1} \,,$$

$$|\partial_x \partial_y z| \approx \frac{1}{2} \sum_{i=1,2} |\partial_y \tilde{z}_h(x_T) - \partial_y \tilde{z}_h(x_i)| \, |x_T - x_i|^{-1} \,,$$

and correspondingly $\partial_y \partial_x z$ and $\partial_y^2 z$ are approximated by changing arguments. In the situation of hanging nodes, see Figure 3.2 (right), constant interpolation to a fictitious midpoint x_1 appears to be sufficiently accurate.

For more irregular meshes, we propose to replace the term $\nabla_h^2 \tilde{z}_h(x_T)$ by $\nabla(\mathcal{M}_h \nabla \tilde{z}_h(x_T))$, using the ((bi-)linear) recovered gradient $\mathcal{M}_h \nabla \tilde{z}_h$. This latter approach is more robust and, by superconvergence arguments, guarantees convergence of $\nabla(\mathcal{M}_h \nabla \tilde{z}_h(x_T))$ at least in the interior of regularly refined parts of the mesh.

3.4   Strategies for mesh adaptation

We briefly discuss how a mesh refinement process may be organized on the basis of an *a posteriori* error estimate of the type (3.8). Suppose that some error tolerance TOL and maximum number N_{max} of mesh points are given. The goal is to find a most economical mesh $\mathbb{T}_h$ on which

$$|J(e)| \approx \eta(u_h) = \sum_{T \in \mathbb{T}_h} \eta_T \approx TOL, \qquad (3.13)$$

with the *local error indicators* $\eta_T := \omega_T \rho_T$. Usually, one starts from an initial coarse mesh which is then successively refined according to the following algorithm:

1. Solve the discrete problem on $\mathbb{T}_h$

2. Evaluate the estimator $\eta = \sum_T \eta_T$

3. If $\eta > TOL$: grid $\mathbb{T}_h$ is changed according to η, go to 1

4. end

There are essentially three alternative strategies to realise point 3 of the above scheme.

1. *Error per cell strategy.* The mesh generation aims to equilibrate the local error indicators η_T, by refining (or coarsening) the elements $T \in \mathbb{T}_h$ according to the criterion

$$\eta_T \approx \frac{TOL}{N}, \qquad N = \#\{T \in \mathbb{T}_h\}. \qquad (3.14)$$

Since, N depends on the result of the refinement decision, this strategy is implicit and would need iteration. However, it is common practice to work with a varying value N on each refinement level which is permanently updated according to the refinement process.

2. *Fixed fraction strategy.* In each refinement cycle, the elements are ordered according to the size of η_T and either a fixed portion (say 30%) of the elements with largest η_T or the portion of elements which make up for a certain part of the estimator, $\kappa\eta(u_h)$, is refined. The appropriate choice of the parameter κ is crucial and depends very much on the particular situation. For "regular" functionals, one may choose $\kappa = 0.6 - 0.8$, while for "singular" functions a smaller choice $\kappa = 0.1 - 0.2$ is advisable, in order to enhance local refinement. This process is repeated until the stopping criterion $\eta(u_h) \approx TOL$ is fulfilled, or N_{max} is exceeded.

3. *Fixed reduction strategy.* One works with a varying tolerance TOL_{var}. If on a mesh $\mathbb{T}_h$ a discrete solution u_h has been obtained with corresponding error estimator $\eta(u_h)$, the tolerance is set to $TOL_{var} = \sigma\eta(u_h)$, with some fixed reduction factor $\sigma \in (0,1)$ (say $\sigma = 0.5$). In the next step, one (or more) cycles of the *error per cell strategy* are applied with tolerance TOL_{var} yielding a refined mesh $\mathbb{T}_h^{new}$ and the new solution u_h^{new} with corresponding error estimator $\eta(u_h^{new})$. Then, the tolerance is reduced again and a new refinement cycle begins. This process is repeated until $TOL_{var} \leq TOL$, or N_{max} is exceeded.

Our experience is, that the first strategy is useful to achieve a grid with $\eta < TOL$ in a few steps. The second refinement strategy is augmented with the feature of coarsening a fixed portion of cells. We use this additional possibility for our time dependent problems to keep about a constant number of cells. Corresponding results are presented in Chapter 13. The third adaptive algorithm is used in situations where we have to work with a regularised linear functional for our adaptive approach. In this case we couple the parameter of regularisation to the value TOL of the adaptive process.

3.5   Example

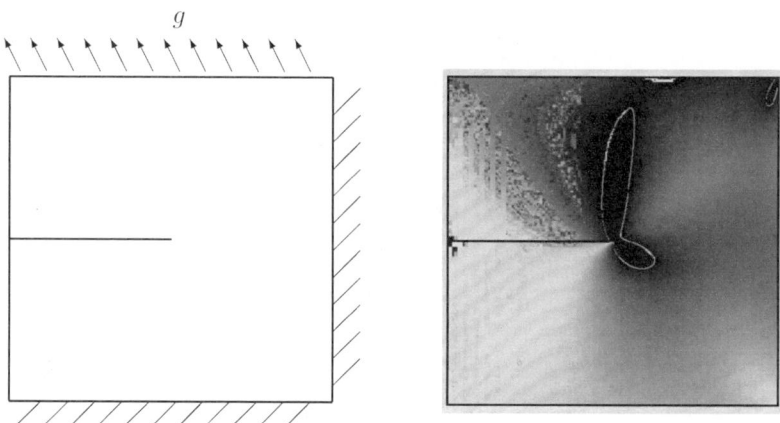

Figure 3.3: Geometry of the square disc test problem and plot of $|\sigma^D|$ (plastic regions black) computed on a mesh with $N \approx 64,000$ cells.

In order to demonstrate the mechanism and benefit of the DWR-method, we show computational results for the following Hencky type problem (for more details of the underlying model, we refer to Chapter 13).

Our test case (cf. Rannacher & Suttmeier [56]) is the plane strain model of a square elasto-plastic disc with a crack subjected to boundary traction acting along the upper boundary (see Figure 3.3). Along the right-hand side and the lower boundary the body is fixed and the remaining part of the boundary (including the crack) is left free. This problem is interesting as its solution develops a singularity at the tip of the crack where in the elastic case a stress concentration occurs with an asymptotic behaviour (expressed in terms of polar coordinates) of the form $\sigma \approx r^{-1/2}$. Hence, local plastification will occur right from the beginning of the loading process.

We consider the case of the linear isotropic material law in the form

$$\sigma = 2\mu\varepsilon(u)^D + \kappa \operatorname{div} u,$$

and the perfect plastic behaviour is determined by the flow function

$$\mathcal{F}(\sigma) = |\sigma^D| - \sqrt{\tfrac{2}{3}}\,\sigma_0 \leq 0.$$

We use the particular values $\kappa = 164206$, $\mu = 80193.80$, $\sigma_0 = 450$, which correspond to Aluminium. The boundary traction is assumed in the form $g = (-111.7, 223.4)$ and we choose $\bar{\gamma} = 0$.

In order to test the behaviour of the adaptive approach described above, we want to evaluate the vertical gap of the crack as our quantity of interest. Our weighted error estimator turns out to be rather sharp even on relatively coarse meshes, see Figure 3.4, left. This indicates that the strategy of evaluating the weights ω_T computationally works also for the present nonlinear problem. Further, this approach yields more economical meshes than the ZZ-indicator, Figure 3.4, right.

The corresponding meshes are shown in Figure 3.5. The *weighted* approach balances the local mesh size around the plastic zone, the corner singularity and the quantity of interest (Figure 3.5,left). The conventional strategy (ZZ,Figure 3.5,right), only resolves the corner singularity.

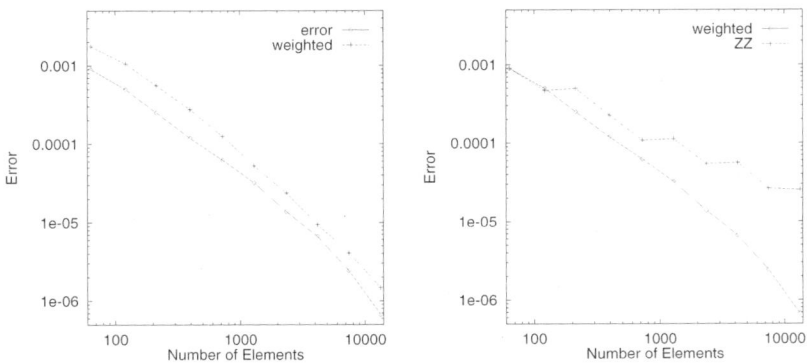

Figure 3.4: On the left, comparison between $|J(\{u_h - u_{ref}, p_h - p_{ref}\})|$ and η_{weight} demonstrating η_{weight} to give sharp estimates. On the right, relative errors for the value of the vertical gap obtained on grids, adaptively refined according to η_{weight} and η_{ZZ}, demonstrating η_{weight} to be more economical.

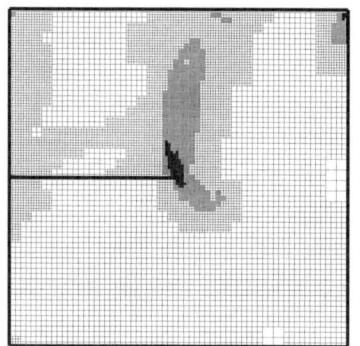 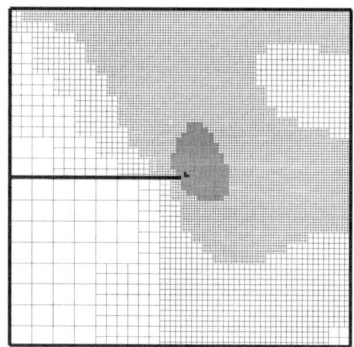

Figure 3.5: Adaptively refined meshes for the test example: The *weighted* approach balances the local mesh size around the plastic zone, the corner singularity and the quantity of interest (3.5,left). The conventional strategy (ZZ,Figure 3.5,right), only resolves the corner singularity.

Chapter 4

Extensions to stabilised schemes

For applying a finite element method to a given problem, it has to be written in a variational setting, which is accessible for a Galerkin scheme. Additional constraints, as for example the incompressibility condition in the Stokes problem, have to be considered as restrictions. These can be treated by the Lagrangian formalism yielding saddle point problems. One important application of such *mixed systems* and the corresponding finite element schemes is the following: For higher order problems one defines auxiliary variables for the derivatives. The equations describing these new quantities are handled as restrictions in the reformulated problem. In this way certain derivatives represented by auxiliary variables are approximated with higher accuracy. On the other hand the new formulation often allows to weaken the regularity assumptions on the primary solution. As one example we mention problems in perfect plasticity. Here, discontinuities may occur in the displacement field u due to slip lines in the micro-structure, whereas the stresses σ, determined by linear combinations of derivatives of u, have a smoother behaviour (see Suquet [65], Seregin [60]). From a practical point of view, the stresses are required with high accuracy. Therefore mixed methods treating the stresses directly are often more adequate for solving problems in continuum mechanics.

In this chapter, we consider the following model situation: The displacement of a thin linear elastic membrane fixed to its boundary and subjected to a volume force f is described by a scalar function u on a domain $\Omega \subset \mathbb{R}^2$, such that

$$- \operatorname{div}\{a(x)\nabla u\} = f \text{ in } \Omega, \qquad u = 0 \text{ on } \partial\Omega,$$

where $\alpha^{-1} \geq a(x) \geq \gamma > 0, x \in \Omega$, describes the local material behaviour. Introducing the stress vector $\sigma = (\sigma_1, \sigma_2)$ the problem can be stated in form of a first order system,

$$\sigma = a(x)\nabla u, \quad -\operatorname{div}\sigma = f \quad \text{in } \Omega, \quad u = 0 \text{ on } \partial\Omega. \tag{4.1}$$

Setting $A = a^{-1}(x) \geq \alpha > 0$, the dual-mixed variational formulation of (4.1) reads

$$(A\sigma, \tau) + (u, \operatorname{div}\tau) - (\operatorname{div}\sigma, \varphi) = (f, \varphi) \quad \forall\{\tau, \varphi\} \in H_{\operatorname{div}}(\Omega) \times L^2(\Omega), \tag{4.2}$$

where $H_{\operatorname{div}}(\Omega) = \{\tau \in L^2(\Omega)^2, \operatorname{div}\tau \in L^2(\Omega)\}$.

Problem (4.2) has a unique solution $\{\sigma, u\} \in H_{\operatorname{div}}(\Omega) \times L^2(\Omega)$ if, and only if the stability conditions

$$(A\tau, \tau) \geq \alpha\|\tau\|_{\operatorname{div}}^2, \quad \tau \in \mathcal{N},$$

$$\sup_{\tau \in H_{\operatorname{div}}} \frac{(\operatorname{div}\tau, \varphi)}{\|\tau\|_{\operatorname{div}}} \geq \beta\|\varphi\|, \quad \varphi \in L^2(\Omega), \tag{4.3}$$

are fulfilled with $\alpha, \beta > 0$, where

$$\mathcal{N} := \{\tau \in H_{\operatorname{div}}, \operatorname{div}\tau = 0 \text{ in } L^2(\Omega)\},$$

$$\|.\|_{\operatorname{div}}^2 := \|.\|^2 + \|\operatorname{div}(.)\|^2.$$

Formally, using the notation $\Sigma_h \subset H_{\operatorname{div}}$ and $V_h \subset L^2(\Omega)$ for the corresponding finite element subspaces, the approximate solution $(\sigma_h, u_h) \in \Sigma_h \times V_h$ of (4.2) is determined by

$$(A\sigma_h, \tau_h) + (u_h, \operatorname{div}\tau_h) - (\operatorname{div}\sigma_h, \varphi) = (f, \varphi_h) \quad \forall\{\tau_h, \varphi_h\} \in V_h \times \Sigma_h. \tag{4.4}$$

Suppose that the discrete analogues of the stability conditions (4.3) are satisfied uniformly with respect to h,

$$(A\tau_h, \tau_h) \geq \tilde{\alpha}\|\tau_h\|_{\operatorname{div}}^2, \quad \tau_h \in \mathcal{N}_h,$$

$$\sup_{\tau_h \in \Sigma_h} \frac{(\operatorname{div}\tau_h, \varphi_h)}{\|\tau_h\|_{\operatorname{div}}} \geq \tilde{\beta}\|\varphi_h\|, \quad \varphi_h \in V_h, \tag{4.5}$$

where $\mathcal{N}_h := \{\tau_h \in \Sigma_h, \operatorname{div} \tau_h = 0 \text{ in } V_h\}$. Then, problem (4.4) is uniquely solvable and the approximation is optimal in the sense that

$$\|\sigma - \sigma_h\|_{\operatorname{div}} + \|u - u_h\| \le C \left\{ \inf_{\tau_h \in \Sigma_h} \|\sigma - \tau_h\|_{\operatorname{div}} + \inf_{\varphi_h \in V_h} \|u - \varphi_h\| \right\},$$

with a generic constant C depending on $\tilde{\alpha}$ and $\tilde{\beta}$. However, it may be viewed as a drawback of this approach that one has to *construct* suitable non-standard finite element spaces satisfying the stability conditions (4.5).

4.1 Discretisation for the membrane-problem

A more straightforward approach, allowing the use of standard finite-element spaces, is based on the concept of least-squares stabilisation as proposed in Hughes et al. [70] and analysed in Franca & Stenberg [32] (see Rannacher & Suttmeier [55] for an application in elasticity). Here the variational formulation (4.2) is augmented by least-squares-like terms based on (4.1) yielding

$$- (A\sigma - \nabla u, \delta_1(A\tau + \nabla\varphi)) + (\operatorname{div}\sigma + f, \delta_2 \operatorname{div}\tau) + (A\sigma, \tau)$$
$$+ (u, \operatorname{div}\tau) - (\operatorname{div}\sigma, \varphi) = (f, \varphi) \qquad \forall \{\tau, \varphi\} \in H_{\operatorname{div}}(\Omega) \times L^2(\Omega), \quad (4.6)$$

where δ_1 and δ_2 are piecewise constant (positive) parameter functions. Defining

$$A_\delta(\{\sigma, u\}, \{\tau, \varphi\}) := (A\sigma, \tau) + (u, \operatorname{div}\tau) - (\operatorname{div}\sigma, \varphi)$$
$$- (A\sigma - \nabla u, \delta_1(A\tau + \nabla\varphi)) + (\operatorname{div}\sigma, \delta_2 \operatorname{div}\tau),$$
$$F_\delta(\{\tau, \varphi\}) := (f, \varphi) - (f, \delta_2 \operatorname{div}\tau),$$

for pairs $\{\tau, \varphi\} \in H_{\operatorname{div}}(\Omega) \times L^2(\Omega)$, the stabilised dual-mixed scheme is described by

$$A_\delta(\{\sigma_h, u_h\}, \{\tau_h, \varphi_h\}) = F_\delta(\{\tau_h, \varphi_h\}) \qquad \forall \{\tau_h, \varphi_h\} \in \Sigma_h \times V_h. \quad (4.7)$$

The natural norm corresponding to the form $A_\delta(.,.)$ is given by

$$\|\{\tau, \varphi\}\|_\delta^2 := (A\tau, \tau) + \|\delta_1^{1/2} \nabla\varphi\|^2 + \|\delta_2^{1/2} \operatorname{div}\tau\|^2.$$

Recalling the assumptions $A = a^{-1}(x)$ and $a(x) \ge \gamma > 0$, the following result ensures unique solvability of the discrete problem (4.7).

Lemma 4.1.1. *For $0 < \delta_1 \leq \frac{1}{2}\gamma$, the bilinear form $A_\delta(.,.)$ is positive definite, i.e.,*

$$A_\delta(\{\tau, \varphi\}, \{\tau, \varphi\}) \geq \frac{1}{2}\|\{\tau, \varphi\}\|_\delta^2 \qquad \forall \{\tau, \varphi\} \in H_{\mathrm{div}}(\Omega) \times L^2(\Omega).$$

Proof: By definition, there holds

$$A_\delta(\{\tau, \varphi\}, \{\tau, \varphi\}) = (A\tau, \tau) - (A\tau, \delta_1 A\tau) + \|\delta_1^{1/2}\nabla\varphi\|^2 + \|\delta_2^{1/2}\,\mathrm{div}\,\tau\|^2.$$

In view of the assumptions for δ_1, the estimate

$$(A\tau, \delta_1 A\tau) \leq \frac{\max_\Omega \delta_1}{\gamma}(A\tau, \tau) \leq \frac{1}{2}(A\tau, \tau)$$

is valid and with this result the assertion follows.

An *a priori* error estimate for the general case of linear elasticity is proven in Rannacher & Suttmeier [55]. We apply this result to the membrane problem.

Theorem 4.1.1. *Let $\Sigma_h \times V_h \subset H_{\mathrm{div}}(\Omega) \times L^2(\Omega)$ be an arbitrary pair of finite-element subspaces. Then the stabilised dual-mixed problem (4.7), with $0 < \delta_1 \leq \frac{1}{2}\gamma$, has an unique solution $\{\sigma_h, u_h\} \in \Sigma_h \times V_h$, and for the error $\{e_\sigma, e_u\} := \{\sigma - \sigma_h, u - u_h\}$ there holds the estimate*

$$\|\{e_\sigma, e_u\}\|_\delta \leq \inf_{\{\tau_h, \varphi_h\} \in \Sigma_h \times V_h} \left\{ 2\|\{\sigma - \tau_h, u - \varphi_h\}\|_\delta + \|\delta_2^{-1/2}(u - \varphi_h)\| + \|\delta_1^{-1/2}(\sigma - \tau_h)\| \right\}.$$

If problem (4.2) is sufficiently regular, such that the usual L^2/H^2-shift theorem holds, then

$$\|e_u\| \leq C \max_\Omega \{h + \delta_2^{1/2} + h\delta_1^{-1/2} + h^2\delta_2^{-1/2}\}\|\{e_\sigma, e_u\}\|_\delta.$$

Remark. For our numerical tests we choose the simplest low-order approximation using (continuous) bilinear shape functions for both unknowns σ_h and u_h. In this case Theorem 4.1.1 yields the *a priori* error estimate

$$\|\{\sigma - \sigma_h, u - u_h\}\|_\delta \leq$$
$$C \max_\Omega\{h^2 + \delta_2^{1/2}h + \delta_2^{-1/2}h^2\}\|\sigma\|_2 + C\max_\Omega\{h^2 + \delta_1^{1/2}h + \delta_1^{-1/2}h^2\}\|u\|_2.$$

Hence, the choice $\delta_1 \approx h$ and $\delta_2 \approx h$ appears asymptotically optimal, yielding the *a priori* error estimate

$$\|\{\sigma - \sigma_h, u - u_h\}\|_\delta \leq Ch_{max}^{3/2}\{\|\sigma\|_2 + \|u\|_2\},$$

Cells	$\|\sigma - \sigma_h\|_\delta$	Ratio	$\|\sigma - \sigma_h\|$	Ratio	$\|u - u_h\|$	Ratio
16	1.899231-02	0.00	1.565212-02	0.00	3.085985-03	0.00
64	5.084942-03	3.73	3.524148-03	4.44	6.342370-04	4.86
256	1.489171-03	3.41	7.874093-04	4.47	1.392413-04	4.55
1024	4.748953-04	3.13	1.760648-04	4.47	3.254831-05	4.27
4096	1.600978-04	2.96	3.975101-05	4.42	7.872258-06	4.13
16384	5.543462-05	2.88	9.091715-06	4.37	1.938442-06	4.06

Table 4.1: Convergence behaviour on uniformly refined meshes for the error measured in the energy norm and the L^2-norm of $(\sigma - \sigma_h)$ and $(u - u_h)$.

and, furthermore,

$$\|u - u_h\| \leq Ch_{max}^2 \left\{ \|\sigma\|_2 + \|u\|_2 \right\} .$$

Using super-approximation effects, on uniform meshes the order of approximation for the stresses may be further improved by post-processing to $\mathcal{O}(h_{max}^2)$.

The discretisation (4.7) is checked for the Laplace equation on $\Omega = (0,1) \times (0,1)$ with a right-hand side corresponding to the solution $u = x(x-1)y(y^2-1)$. On a sequence of globally refined meshes, we evaluate the error in the energy norm. Additionally, we compute the L^2-errors for the stress vector σ and the displacement u. The results are presented in Table 4.1. Evaluating the error reduction rate *Ratio*, we observe convergence behaviour of $\mathcal{O}(h^{3/2})$ for $\|\sigma - \sigma_h\|_\delta$. Furthermore, the improved order of $\mathcal{O}(h^2)$ for the error of u and σ measured in the L^2-norm is indicated.

4.2 A posteriori error analysis

We employ techniques developed in Becker & Rannacher [8] and Rannacher & Suttmeier [55] for estimating the error of scheme (4.7) with respect to general measures given in terms of linear functionals J(.) defined on the space $H_{\text{div}} \times L^2(\Omega)$ or appropriate subspaces. Relevant examples are torsion moments, stress values, or mean surface tension

$$J_\psi(\{\sigma, u\}) = \int_\Omega u\psi \, dx, \qquad J_i(\{\sigma, u\}) = \sigma_i(x_0), \qquad J_\Gamma(\{\sigma, u\}) = \int_\Gamma \sigma n \, ds.$$

The derivation of a posteriori error estimates for the stabilised mixed scheme (4.7) requires some care. Following an approach proposed in Rannacher [54], we use the solution $\{\zeta, z\} \in H_{\mathrm{div}} \times L^2(\Omega)$ of the corresponding (stabilised) dual problem

$$A_\delta(\{\tau, \varphi\}, \{\zeta, z\}) = J(\{\tau, \varphi\}) \qquad \forall \{\tau, \varphi\} \in H_{\mathrm{div}} \times L^2(\Omega), \qquad (4.8)$$

to derive a representation for the measure of the error $\{e_\sigma, e_u\}$. This is accomplished as follows.

First, taking the difference between (4.6) and (4.7), we have the orthogonality relation

$$A_\delta(\{e_\sigma, e_u\}, \{\zeta_h, z_h\}) = 0, \qquad \{\zeta_h, z_h\} \in \Sigma_h \times V_h. \qquad (4.9)$$

Now, choosing $\{\tau, \varphi\} = \{e_\sigma, e_u\}$ in (4.8) and exploiting (4.9), we obtain

$$
\begin{aligned}
J(\{e_\sigma, e_u\}) &= A_\delta(\{e_\sigma, e_u\}, \{\zeta - \zeta_h, z - z_h\}) \\
&= (\nabla u_h - A\sigma_h, \zeta - \zeta_h) + (f + \mathrm{div}\,\sigma_h, z - z_h) \\
&\quad - (f + \mathrm{div}\,\sigma_h, \delta_2\,\mathrm{div}(\zeta - \zeta_h)) + (A\sigma_h - \nabla u_h, \delta_1 A(\zeta - \zeta_h)) \\
&\quad + (A\sigma_h - \nabla u_h, \delta_1 \nabla(z - z_h)).
\end{aligned}
$$

From this we conclude the error estimate

$$|J(\{e_\sigma, e_u\})| \leq \sum_{T \in \mathbb{T}_h} \left\{ \omega_T^{(1)} \rho_T^{(1)} + \omega_T^{(2)} \rho_T^{(2)} \right\},$$

with the local residuals $\rho_T^{(i)}$ and weight factors $\omega_T^{(i)}$ defined by

$$
\begin{aligned}
\rho_T^{(1)} &= \|f + \mathrm{div}\,\sigma_h\|_T, & \omega_T^{(1)} &= \|z - z_h\|_T + \delta_{2,T} \|\mathrm{div}(\zeta - \zeta_h)\|_T, \\
\rho_T^{(2)} &= \|A\sigma_h - \nabla u_h\|_T, & \omega_T^{(2)} &= \|\zeta - \zeta_h\|_T + \delta_{1,T} \{ \|\nabla(z - z_h)\|_T \\
& & & \qquad + \|A(\zeta - \zeta_h)\|_T \}.
\end{aligned}
$$

In general, the weights $\omega_T^{(i)}$ cannot be determined analytically, but have to be computed by solving the dual problem numerically on the current mesh. In order to get practically useful bounds for the weights, one uses standard interpolation estimates and approximates the appearing second order derivatives by simply taking second order difference quotients of the approximate dual solution $\{\tilde{\zeta}_h, \tilde{z}_h\} \in \Sigma_h \times V_h$, yielding

$$\omega_T^{(1)} \approx \tilde{\omega}_T^{(1)} := C_i h_T \left\{ h_T^2 |\nabla_h^2 \tilde{z}_h(x_T)| + \delta_{2,T} h_T |\nabla_h^2 \tilde{\zeta}_h(x_T)| \right\}, \tag{4.10}$$

$$\omega_T^{(2)} \approx \tilde{\omega}_T^{(2)} := C_i h_T \left\{ h_T^2 |\nabla_h^2 \tilde{\zeta}_h(x_T)| + \delta_{1,T} h_T |\nabla_h^2 \tilde{z}_h(x_T)| \right\}, \tag{4.11}$$

where x_T is the mid–point of element T, and ∇_h^2 is an appropriate difference operator approximating ∇^2. The interpolation constant C_i may be obtained by calibration on coarser meshes or simply be set to $C_i = 1$. This results in the approximate a posteriori error bound

$$|J(\{e_\sigma, e_u\})| \approx \eta_{weight}^{stab} := C_i \sum_{T \in \mathbb{T}_h} \left\{ \tilde{\omega}_T^{(1)} \rho_T^{(1)} + \tilde{\omega}_T^{(2)} \rho_T^{(2)} \right\}. \tag{4.12}$$

We will compare the performance of the weighted error estimator (4.12) against two more traditional ones.

1) The standard dual approach: Here, in contrast to (4.8), the error representation is based on the standard dual problem without stabilisation

$$(A\tau, \zeta) + (\varphi, \operatorname{div} \zeta) - (\operatorname{div} \tau, z) = J(\{\tau, \varphi\}) \quad \forall \{\tau, \varphi\} \in H_{\operatorname{div}} \times L^2(\Omega). \tag{4.13}$$

As above, choosing $\{\tau, \varphi\} = \{e_\sigma, e_u\}$ in (4.13) and exploiting (4.9), we obtain in this case

$$\begin{aligned}
J(\{e_\sigma, e_u\}) &= (Ae_\sigma, \zeta) + (e_u, \operatorname{div} \zeta) - (\operatorname{div} e_\sigma, z) - A_\delta(\{e_\sigma, e_u\}, \{\zeta_h, z_h\}) \\
&= (\nabla u_h - A\sigma_h, \zeta - \zeta_h) + (f + \operatorname{div} \sigma_h, z - z_h) \\
&\quad + (f + \operatorname{div} \sigma_h, \delta_2 \operatorname{div} \zeta_h) - (A\sigma_h - \nabla u_h, \delta_1 A\zeta_h) \\
&\quad - (A\sigma_h - \nabla u_h, \delta_1 \nabla z_h),
\end{aligned}$$

yielding the error estimate

$$|J(\{e_\sigma, e_u\})| \approx \eta_{weight} := C_i \sum_{T \in \mathbb{T}_h} \left\{ \bar{\omega}_T^{(1)} \rho_T^{(1)} + \bar{\omega}_T^{(2)} \rho_T^{(2)} \right\}, \tag{4.14}$$

where the weights $\bar{\omega}_T^{(i)}$ are determined by

$$\bar{\omega}_T^{(1)} = \|z - z_h\|_T + \delta_{2,T} \|\operatorname{div} \zeta_h\|_T,$$

$$\bar{\omega}_T^{(2)} = \|\zeta - \zeta_h\|_T + \delta_{1,T} \{ \|\nabla z_h\|_T + \|A\zeta_h\|_T \}.$$

2) An energy-error estimator: In order to obtain an energy estimator for the $\|.\|_\delta$-norm, we start with the result of Lemma 4.1.1 and use (4.9) to get

$$\frac{1}{2}\|\{e_\sigma, e_u\}\|_\delta^2 \le A_\delta(\{e_\sigma, e_u\}, \{e_\sigma, e_u\}) = A_\delta(\{e_\sigma, e_u\}, \{e_\sigma, \tilde{e}_u\})$$
$$= (\nabla u_h - A\sigma_h, e_\sigma) + (f + \operatorname{div}\sigma_h, \tilde{e}_u)$$
$$- (f + \operatorname{div}\sigma_h, \delta_2 \operatorname{div} e_\sigma) + (A\sigma_h - \nabla u_h, \delta_1 A e_\sigma)$$
$$+ (A\sigma_h - \nabla u_h, \delta_1 \nabla \tilde{e}_u),$$

where $\tilde{e}_u = u - \tilde{u}_h$ with $\tilde{u}_h$ being a suitable interpolant of u. Now using Young's inequality and the interpolation estimate $\|u - \tilde{u}_h\|_T \le C h_T \|\nabla u\|_{\tilde{T}}$, where $\tilde{T}$ is the patch consisting of T and its neighbours, we obtain the estimate (cf. Rannacher & Suttmeier [55])

$$\|\{e_\sigma, e_u\}\|_\delta^2 \le \eta_{energy}^2 = C \sum_{T \in \mathbb{T}_h} \left\{ h_T \rho_T^{(1)2} + \rho_T^{(2)2} \right\}, \tag{4.15}$$

with constants C and the residual terms $\rho_T^{(i)}$ as defined above.

Remark: The energy estimate (4.15) is tested at the Laplace example described above.

In Figure 4.1 (left) the energy error, the estimated error and its two components

$$\text{rho1} = \left(\sum_{T \in \mathbb{T}_h} h_T \rho_T^{(1)2} \right)^{1/2}, \qquad \text{rho2} = \left(\sum_{T \in \mathbb{T}_h} \rho_T^{(2)2} \right)^{1/2}$$

are depicted. We observe the second part of (4.15) to be of only sub-optimal order. One may account for this deficiency by using the modified residuum $\tilde{\rho}_T^{(2)} = \|P(A\sigma_h - \nabla u_h)\|_T$, where P denotes the L^2-projection onto Σ_h. In Figure 4.1 (right) the correct asymptotic behaviour of the modified energy estimate is shown. The theoretical justification of this process is left as an open problem. In order to check that no modification of $\rho_T^{(2)}$ is required in (4.12), we apply an a posteriori estimate for the energy norm of the form

$$\|\{e_\sigma, e_u\}\|_\delta \le \eta_{weight}^{stab},$$

to the Laplace problem, using the duality techniques described above, where the functional $J(.)$ in (4.8) is taken as

$$J(\{\tau, \varphi\} = \|\{e_\sigma, e_u\}\|_\delta^{-1} A_\delta(\{e_\sigma, e_u\}, \{\tau, \varphi\})).$$

Comparing η_{weight}^{stab} and η_{energy} (see Figure 4.2), this approach yields, in contrast to the coercivity argument, an asymptotically correct bound for the energy error.

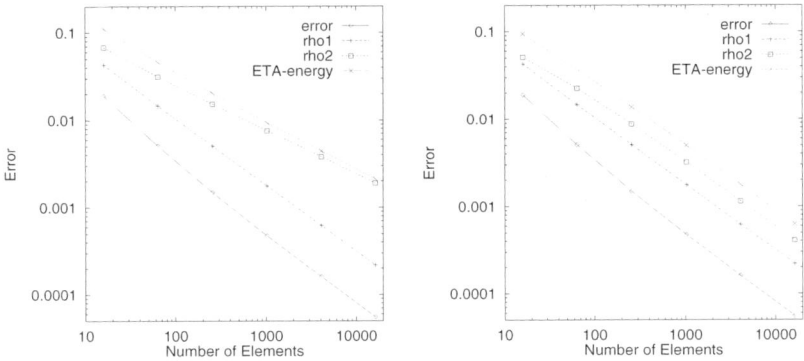

Figure 4.1: Energy error, estimated error η_{energy} (consisting of components rho1 and rho2) with $\rho_T^{(2)}$ (left) and modified $\rho_T^{(2)}$ (right), for the Laplace problem.

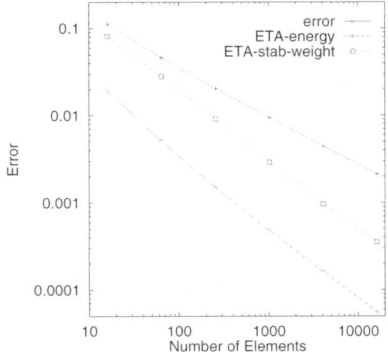

Figure 4.2: Comparison of η_{weight}^{stab} and η_{energy} showing that the duality techniques produce a sharper estimate for $\|\{e_\sigma, e_u\}\|_\delta$.

4.3   Numerical tests

As a numerical test the membrane problem is considered in the classical form

$$- \operatorname{div}\{a(x)\nabla u\} = f \, ,$$

with $f = 1$ on the crack domain depicted in Figure 4.3. The outer rectangle is $[-4;1] \times [-1;1]$, the crack is determined by the points $(0,0)$ and $(1,0)$, and the point P has coordinates $(-3, 0.5)$. Furthermore we choose

$$a(x) = \begin{cases} a_1 > 0 & \text{for } -0.5 < x_1 < 0.5 \, , \ -0.75 < x_2 < 0.75 \, , \\ a_0 > 0 & \text{otherwise} \, . \end{cases}$$

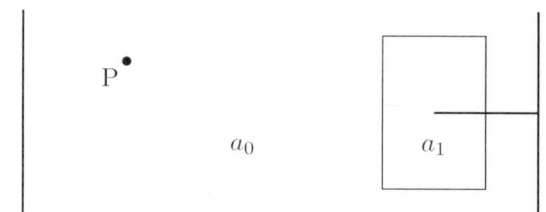

Figure 4.3: Crack domain for the numerical test.

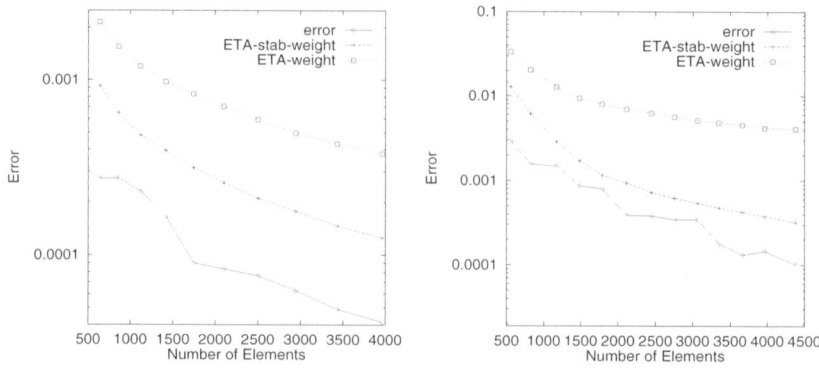

Figure 4.4: Comparison between the error $|J(\sigma_h) - J(\sigma_{ref})|$ and η_{weight}^{stab} and η_{weight}, for the first (left) and the second (right) example, demonstrating η_{weight}^{stab} to give sharper estimates.

We apply our technique for a posteriori error estimation to the evaluation of the stress point value $\sigma_1(P)$, giving the error functional

$$J(\sigma) = J(\{\sigma, u\}) = \sigma_1(P),$$

1) The first numerical test is done with $a_0 = a_1 = 1$. In Table 4.2 (left) and Figure 4.4 (left), the ratio between true and estimated error is investigated. It is shown, that η_{weight}^{stab} yields a sharp estimate, whereas using η_{weight} results in an increasing ratio indicating the wrong asymptotic behaviour.

In Figure 4.5 (left), the relative errors obtained on grids, adaptively refined according to η_{weight}^{stab}, η_{weight} and η_{energy}, are depicted. One observes the first one to be more economical, i.e. higher accuracy is achieved with the same amount of cells, compared to the latter ones.

2) The same effects can be observed in the second numerical test with $a_0 = 1$ and $a_1 = 5$. With no further comments the results are given in Figure 4.5 (right), Table 4.2 (right) and Figure 4.4 (right).

Eventually, the adaptive grids produced on the basis of the three error estimates are shown in Figure 4.6 and Figure 4.7 for the first and second example respectively.

Cells	RelErr	$\text{Rat}_{weight}^{stab}$	Rat_{weight}	Cells	RelErr	$\text{Rat}_{weight}^{stab}$	Rat_{weight}
1120	1.985-3	2.09	3.16	2113	7.370-4	2.27	4.22
1420	1.410-3	2.40	4.27	2446	7.281-4	1.77	3.82
1744	7.716-4	3.51	5.11	2755	6.574-4	1.68	3.83
2104	7.141-4	3.11	6.49	3067	6.574-4	1.47	3.48
2503	6.522-4	2.77	6.66	3352	3.381-4	2.51	6.38
2947	5.379-4	2.83	6.38	3661	2.495-4	3.04	8.11
3433	4.210-4	2.98	7.05	3967	2.746-4	2.46	6.77
3964	3.566-4	3.02	8.13	4379	1.923-4	2.99	9.44

Table 4.2: Comparison between the Ratio for η_{weight}^{stab} and η_{weight}, left table for the first, right table for the second example, demonstrating that only η_{weight}^{stab} has the correct asymptotic behaviour.

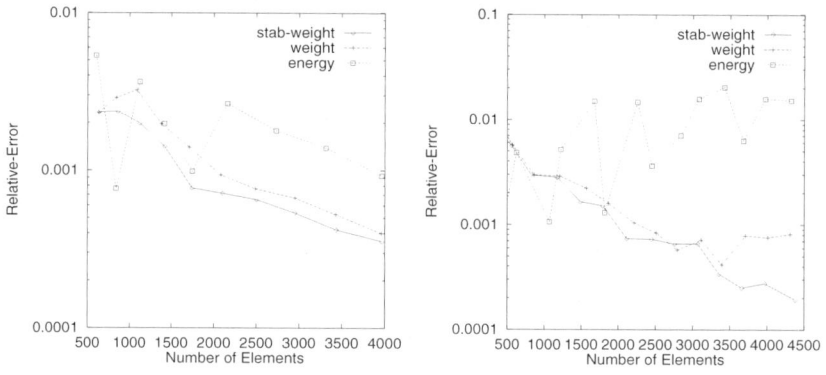

Figure 4.5: Relative errors for the stress point-value $\sigma_1(p)$ based on the different estimators for the first (left) and the second (right) example, demonstrating η^{stab}_{weight} to be most economical.

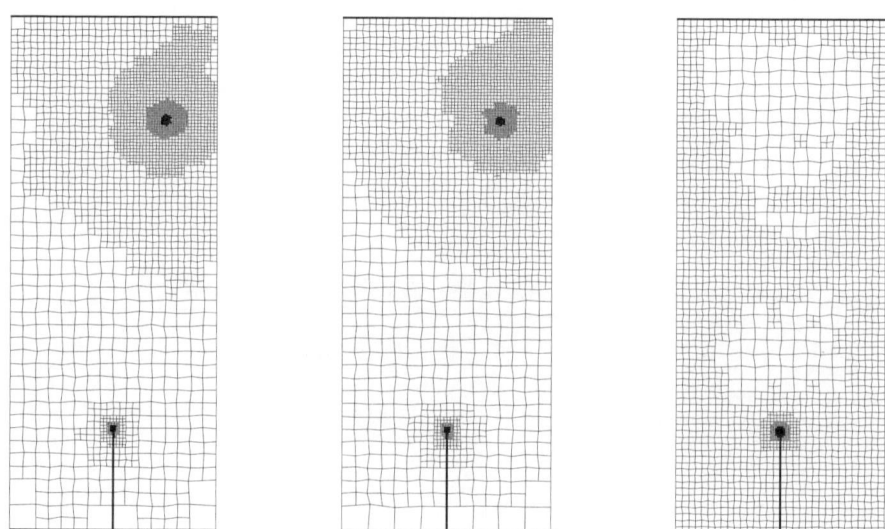

Figure 4.6: Structure of grids, for the first example, based on η^{stab}_{weight} (left), η_{weight}(middle) and η_{energy}(right) with $N \approx 3000$.

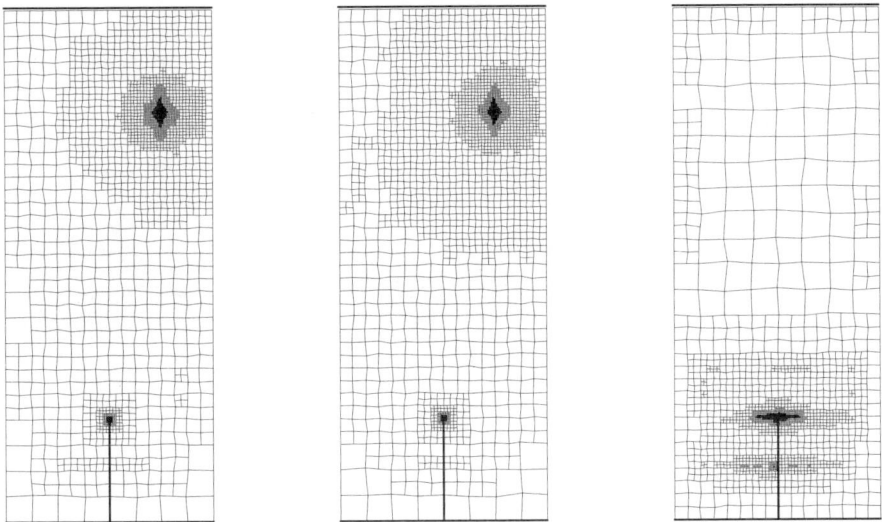

Figure 4.7: Structure of grids, for the second example, based on η_{weight}^{stab} (left), η_{weight}(middle) and η_{energy}(right) with $N \approx 3000$.

Chapter 5

Obstacle problem

As a model example, we choose the so-called obstacle problem (cf. Lions and Stampacchia [48]) to demonstrate, how a problem formulated in form of a variational inequality can be attacked directly to derive *a posteriori* error estimates analogously to the example in the introduction. Our procedure is performed in three steps, following the approach in the context of variational equalities, i.e., *a priori* error analysis is first done for the energy norm, then extended to L^2-norm estimates finally leading to *a posteriori* error bounds.

The classical solution u of this model problem is determined by the system in $\Omega \subset \mathbb{R}^2$

$$
\begin{aligned}
-\Delta u - f &\geq 0\,, \\
u - \psi &\geq 0\,, \\
(u - \psi)(-\Delta u - f) &= 0\,,
\end{aligned}
\tag{5.1}
$$

and subjected to the homogeneous boundary condition $u = 0$ on $\partial \Omega$.

The basis for applying the finite element method to (5.1) is the formulation as a variational inequality, which takes the form

$$
(\nabla u, \nabla(\varphi - u)) \geq (f, \varphi - u) \qquad \forall \varphi \in K,
\tag{5.2}
$$

where we set $V = \{v \in H^1(\Omega) \mid v = 0 \text{ on } \partial \Omega\}$ and $K = \{v \in V \mid v \geq \psi \text{ in } \Omega\}$, with the obstacle $\psi : \Omega \to \mathbb{R}$. In what follows, ψ_h denotes the piecewise bilinear interpolant of ψ with respect to a quadrilateral FE-mesh.

Equation (5.2) is uniquely solvable (cf. Lions and Stampacchia [48]) and under appropriate smoothness conditions on the boundary and data, the solution is known to satisfy the regularity result $u \in H^2(\Omega)$ (see Brézis and Stampacchia [18]).

The finite element approximation u_h of u in (5.2) is determined by

$$(\nabla u_h, \nabla(\varphi - u_h)) \geq (f, \varphi - u_h) \qquad \forall \varphi \in K_h, \tag{5.3}$$

where $K_h = \{v \in V_h \mid v \geq \psi_h \text{ in } \Omega\}$. The finite dimensional problem can be shown to be uniquely solvable following the same line of arguments as in the continuous case.

5.1   Energy norm

First, we prepare the derivation of *a priori* error estimates in the energy norm.

Setting $\delta = \psi - \psi_h$, the functions $\varphi = u_h + \delta$ and $I_h u \in K_h$, where I_h denotes standard interpolation operator, are admissible testfunctions in (5.2) and (5.3) respectively. So we obtain the modified Galerkin estimate

$$
\begin{aligned}
(\nabla(u - u_h), &\nabla(I_h e - \delta)) \\
&= -(f, \delta) + \underbrace{(f, I_h e) - (\nabla u_h, \nabla(I_h e - \delta))}_{\leq 0 \text{ use (5.3)}} + (\nabla u, \nabla(I_h e - \delta) - \nabla(e - \delta))) \\
&\quad - (f, (I_h e - \delta) - (e - \delta))) + \underbrace{(\nabla u, \nabla(e - \delta)) - (f, e - \delta)}_{\leq 0 \text{ use (5.2)}} \\
&\leq (\nabla u, \nabla(I_h e - e)) - (f, I_h e - e) - (f, \delta) - (\nabla u_h, \nabla \delta).
\end{aligned}
$$

After these preparations the announced estimate (cf. e.g., Falk [30] and Brezzi et al. [19]) is obtained by

$$
\begin{aligned}
\|\nabla e\|^2 &= (\nabla(u - u_h), \nabla(e - I_h e)) + (\nabla(u - u_h), \nabla(I_h e - \delta)) + (\nabla(u - u_h), \nabla \delta) \\
&\leq (\nabla(u - u_h), \nabla(e - I_h e)) + (\nabla u, \nabla(I_h e - e)) \\
&\quad - (f, I_h e - e) + (\nabla u, \nabla \delta) - (f, \nabla \delta) \\
&\leq \frac{1}{2}\|\nabla(u - u_h)\|^2 + \frac{1}{2}\|\nabla(u - I_h u)\|^2 + \|\Delta u + f\|\left\{\|u - I_h u\| + \|\psi - \psi_h\|\right\},
\end{aligned}
$$

yielding

Theorem 5.1.1. *For scheme* (5.3), *there holds the* a priori *error estimate in the energy norm*

$$\|\nabla(u - u_h)\|^2 \leq c\mathbf{h^2}\|\nabla^2 u\|^2 + c\|\Delta u + f\|\mathbf{h^2}\Big\{\|\nabla^2 u\| + \|\nabla^2 \psi\|\Big\}.$$

5.2 Duality argument

In order to establish an improved *a priori* estimate for the discretisation error measured in the L^2-Norm, motivated by Natterer [51] we would like to define a dual solution z for a given error functional J by a variational inequality, searching for $z \in G$ such that

$$(\nabla(\varphi - z), \nabla z) \geq J(\varphi - z) \qquad \forall \varphi \in G, \tag{5.4}$$

with $J(\cdot) = (e, .)$. Here, $G \subset V$ is a closed convex subset which will be appropriately chosen below. The choice $\varphi = z + e = z + (u - u_h)$ and further estimation of the left hand side of (5.4) would then yield the desired estimate for $J(e)$.

Recalling that we shall not require full conformity of the discretisation, $K_h = K \cap V_h$, i.e., we allow $\delta = \psi - \psi_h$ to be non zero. To handle this case it turns out to be convenient to modify (5.4), defining $z \in G$ by

$$(\nabla(\varphi - z), \nabla z) \geq J((\varphi + \delta) - z) \qquad \forall \varphi \in G. \tag{5.5}$$

In order to get an estimate for $J(e)$ we have to require that G is chosen such that $\theta = z + e + \delta$ is again an element of G. Using θ as a test function in (5.5) we immediately get

$$J(e) \leq -(\nabla\delta, \nabla z) + (\nabla(u - u_h), \nabla z)$$
$$= -(\nabla\delta, \nabla z) + (\nabla(u - u_h), \nabla(z - z_h)) + (\nabla(u - u_h), \nabla z_h). \tag{5.6}$$

In the case of variational equalities the last term would vanish identically by the standard Galerkin orthogonality. In the second term on the right we could exploit the result of the energy error estimate and stability properties of the dual problem leading to *a priori* estimates. To proceed along the same line of argument, we insert the right-hand side of the problem into (5.6), to obtain

$$J(e) \leq -(\nabla\delta, \nabla z) + (\nabla(u - u_h), \nabla(z - z_h)) + (\nabla u, \nabla(z_h + u_h + \delta - u))$$
$$- (f, z_h + u_h + \delta - u) + \underbrace{(f, z_h) - (\nabla u_h, \nabla z_h)}_{I}$$
$$+ \underbrace{(\nabla u, \nabla(u - (u_h + \delta))) - (f, u - (u_h + \delta))}_{II} . \quad (5.7)$$

We shall discuss below how the subsets G can be chosen in such a way that the sum of the last two terms becomes non positive and finally there holds

$$J(e) \leq -(\nabla\delta, \nabla z) + (\nabla(u - u_h), \nabla(z - z_h)) + (\nabla u, \nabla(z_h - z)) - (f, z_h - z) .$$

Eventually, assuming a stability result of the form $\|\nabla^2 z\| \leq C_S h^{-\alpha} \|e\|$, with some $\alpha \geq 0$ we can estimate $\|e\|^2 \leq \left\{ h^2 \|\nabla^2 \psi\| + h\|\nabla(u - u_h)\| + h^2 \|\Delta u + f\| \right\} \|\nabla^2 z\|$, yielding

Theorem 5.2.1. *Under the assumption of a stability result of the form* $\|\nabla^2 z\| \leq C_S h^{-\alpha} \|e\|$, *with some* $\alpha \geq 0$, *the* L^2-*error for scheme* (5.3), *can be estimated by*

$$\|e\| \leq C C_S h^{-\alpha} h^2 ,$$

with $C = C(\|\nabla^2 \psi\|, \|\nabla^2 u\|, \|\Delta u + f\|)$.

It remains the discussion of (5.7) and the choice of G. As above, observing that $\varphi = u_h + \delta \in K$, we see from (5.2) that term II in (5.7) is non positive.

Next, we use the discrete contact set $B_h = \{x \in \Omega \mid u_h(x) = \psi_h\}$ and set $W_h^{\psi} = \{v \in V \mid v \geq \psi_h \text{ on } B_h\} \cap V_h$ and W_h^0 correspondingly, using $\psi_h = 0$. Defining $\tilde{u}_h \in W_h^{\psi}$ by the discrete variational inequality

$$(f, \varphi - \tilde{u}_h) - (\nabla(\tilde{u}_h), \nabla(\varphi - \tilde{u}_h)) \leq 0 \qquad \forall \varphi \in W_h^{\psi} , \quad (5.8)$$

we immediately see that $\tilde{u}_h = u_h$, since we only have removed non-active constraints in (5.3). Thus, if we require $z_h \in W_h^0$, i.e. $z_h \geq 0$ on B_h, and choose $\varphi = u_h + z_h \in W_h^{\psi}$ in (5.8) we see that the term I on the right of (5.7) is non positive.

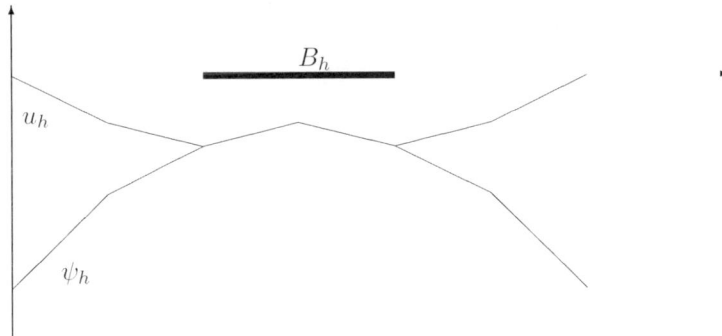

Figure 5.1: Sketch for illustrating the construction of W_h.

The preceeding considerations lead to a natural choice for the convex subsets G, used to define the dual problem,

$$G = \{v \in V \mid v \geq 0 \text{ on } B_h$$
$$\text{and } (f, v + u_h + \delta - u) - (\nabla u, \nabla(v + u_h + \delta - u)) \geq 0\}.$$

It remains to check whether the test function $\theta = z + u - u_h + \delta$ used in (5.5) really belongs to G. This, however, is trivial since in B_h we have $u \geq \psi$, $u_h = \psi_h$, and this gives us $z + u - (u_h + \delta) \geq 0$ in B_h. Thus θ fulfils the first condition in the definition of G.

Furthermore, with the choice $\varphi = u_h \in K$ in (5.2) and because $z \in G$, there holds

$$0 \leq (f, \nabla(u - u_h)) - (\nabla u, \nabla(u - u_h)) + (f, z + u_h - u) - (\nabla u, \nabla(z + u_h - u))$$
$$= (f, (z + u - u_h)) + u_h - u)) - (\nabla u, \nabla((z + u - u_h)) + u_h - u)).$$

This means that θ fulfils the second condition in the definition of G, which completes the proof.

5.3   A posteriori estimates

In order to establish *a posteriori* estimates for the discretisation error measured in form of a (linear) functional $J(\cdot)$, we employ the duality argument from the subsection above.

As a starting point we insert f into the estimate (5.6) yielding

$$
\begin{aligned}
J(e) \leq\ &-(\nabla\delta, \nabla z) + (f, z - z_h) - (\nabla u_h, \nabla(z - z_h)) \\
&+ (\nabla u, \nabla(z - z_h)) - (f, z - z_h) + (\nabla(u - u_h)\nabla z_h)\,.
\end{aligned} \quad (5.9)
$$

The last three terms can be identically transformed into

$$
\begin{aligned}
(\nabla u, \nabla(z - z_h)) &- (f, z - z_h) + (\nabla(u - u_h), \nabla z_h) \\
= (f, z_h) - (\nabla u_h, \nabla z_h) &+ (\nabla u, \nabla(z + u_h + \delta - u)) - (f, z + u_h + \delta - u) \\
&+ (\nabla u, \nabla(u - (u_h + \delta))) - (f, u - (u_h + \delta))\,.
\end{aligned}
$$

Recalling the arguments from the preceeding subsection the right hand side can be estimated from above by zero. Eventually one gets

$$
J(e) \leq -(\nabla\delta, \nabla z) + (f, z - z_h) - (\nabla u_h, \nabla(z - z_h))\,.
$$

Cell-wise integration by parts results in

Theorem 5.3.1. *For the scheme* (5.3) *there holds the a posteriori error bound*

$$
|J(e)| \leq \sum_{T \in \mathbb{T}_h} \omega_T \rho_T + \sum_{T \in \mathbb{T}_h} \Psi_T\,, \quad (5.10)
$$

with $\Psi_T = -(\nabla\delta, \nabla z)_T$ *and local residuals* ρ_T *and weights* ω_T *defined by*

$$
\rho_T := h_T \|f + \Delta u_h\|_T + \tfrac{1}{2} h_T^{1/2} \|n \cdot [\nabla u_h]\|_{\partial T}\,,
$$
$$
\omega_T := \max \left\{ h_T^{-1} \|z - z_h\|_T, h_T^{-1/2} \|z - z_h\|_{\partial T} \right\}\,,
$$

where for interior interelement boundaries $[\partial_n u_h]$ *denotes the jump of the normal derivative* $\partial_n u_h$*.*

In general, the weights ω_T cannot be determined analytically, but have to be computed by solving the dual problem numerically on the available mesh. Next, one uses the interpolation estimates

$$
\omega_T \leq C_{i,T} h_T \|\nabla^2 z\|_T\,, \quad (5.11)
$$

for $z \in H^2(T)$.

An approximation $\tilde{z}_h$ of the dual problem (5.5) may be obtained as follows. The difference $e = u - u_h$ can be approximated by $\tilde{e}$ obtained by extrapolation techniques. Consequently, the modified set is

$$\tilde{G} = \{v \in V \mid v \geq 0 \text{ on } B_h \text{ and } (f, v + \delta - \tilde{e}) - (\sigma_h, \nabla(v + \delta - \tilde{e})) \geq 0\}.$$

A more heuristic idea is to employ *a posteriori* error estimates with respect to the linear Dirichlet problem

$$(\nabla v, \nabla \varphi) = (f, \varphi) \qquad \text{on } \Omega \setminus B_h.$$

This can be justified using our approach as follows. First, we assume that the contact zone is already sufficiently well resolved, i.e., $B_h \approx B = \{x \in \Omega \mid u(x) = \psi(x)\}$. Further, we assume that there holds the strict complementarity condition in (5.1), i.e., $-\Delta u - f > 0$ on B_h and $-\Delta u - f = 0$ on $\Omega \setminus B_h$. The second condition in the definition of G can therefore be approximated as follows

$$0 \leq (f, v + u_h + \delta - u) - (\nabla u, \nabla(v + u_h + \delta - u))$$
$$\approx \int_{B_h} (f + \Delta u)(v + u_h + \delta - u)\, dx \approx \int_{B_h} (f + \Delta u) v\, dx.$$

This would imply that $v \leq 0$ on B_h. Together with the first condition in G we might thus approximate G by $\tilde{G} = \{v \in V \mid v = 0 \text{ on } B_h\}$.

Therefore, we only have to solve a linear Dirichlet problem on $\Omega \setminus B_h$ with zero boundary conditions yielding a discrete dual solution $\tilde{z}_h$. This can be employed to make the approximation

$$\omega_T \approx \tilde{\omega}_T := \tilde{C}_{i,T} h_T^2 |\nabla_h^2 \tilde{z}_h(x_T)|.$$

and one evaluates the right hand side by simply taking second order difference quotients of the approximate dual solution $\tilde{z}_h \in V_h$, where x_T is the mid–point of element T.

Warning: Our last heuristic strategy may fail, which we illustrate by the following example in one space dimension.

Consider an obstacle problem on $\Omega \subset \mathbb{R}$ where ψ is chosen as the solution of the unrestricted problem. In this case, the corresponding FE-solution is simply the interpolant of ψ. Consequently there holds $B_h = \Omega$, i.e. we have $\tilde{G} = \{v \in V | v = 0 \text{ on } \Omega\}$ implying the dual solution z to be zero on Ω. Using the estimate (5.10) would suggest $|J(e)| \leq 0$, which is not true for arbitraty $J(\cdot)$. A remedy is given below in Chapter 10.

Remark: If one is seeking control on norms instead of linear functionals of the error, e.g., L^2-error $\|e\|$ or L^2-norm $\|u\|$ (see the numerical test below), the corresponding functionals may be taken as

$$J_e(\varphi) = (u - u_h, \varphi)/\|u - u_h\|, \qquad J_u(\varphi) = (u + u_h, \varphi). \tag{5.12}$$

In both cases the evaluation of (5.10) at first requires the generation of initial guesses of $u \pm u_h$ from the preceding computations on coarser meshes. Results concerning the reliability of estimating J in this way can be found, e.g., in Backes [4].

Remark: Other approaches to local a posteriori error estimators for variational inequalities of the type discussed in this chapter are presented e.g. in Ainsworth et al. [2], Kornhuber [45], Veeser [71]. Here the error is estimated in the energy norm only. The strategies are mainly based on generalisations of the element residual methods of Bank & Weiser [6].

5.4 Numerical results

As a first test example, we consider (5.1) on $\Omega = (0, 1)^2$ with $f = 0$. The obstacle is given by $\psi(x) = \exp\left((y^2 - 1)^{-1}\right)$ with $y = (x - x_m)/r$, $x_m = (0.5, 0.5)$, $r = 0.5$. We want to control $\|u\|^2$ in an area B_C around the boundary of the contact zone, where approximately $B_C \approx \{x \in \Omega \mid 0.24 < |x| < 0.31\}$ and the functional J is the obvious modification of J_u in (5.12). The computational results are shown in Table 5.1,(left). Evaluating *Ratio* shows the constant relation between the *true* error and the corresponding estimation on sufficiently fine grids, and consequently it is demonstrated that the proposed approach to a *posteriori* error control gives useful error bounds. An adaptive grid, generated on the basis of the weighted estimate, is depicted in Figure 5.2.

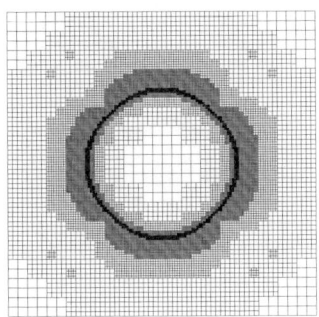

Figure 5.2: On the basis of the weighted estimate adaptively refined grid for the first test-example.

As a second test, we consider (5.1) with $f = 0$ on a slit domain $\Omega \subset (0,1)^2$ where the slit is determined by $\{x \in \mathbb{R}^2 \mid x_1 < 0.5 \text{ and } x_2 = 0.5\}$. The obstacle is given by $\psi(x) = \exp\left((y^2 - 1)^{-1}\right)$, with $y = (x - x_m)/r$, $x_m = (0.25, 0.75)$, $r = 0.2$. For this test, we choose $J(\varphi) = \varphi(x_0)$, $x_0 = (1/3, 1/3)$, to control the point-error in x_0. The numerical results are presented in Table 5.1,(right). Again, it is demonstrated, that the proposed approach to a *posteriori* error control gives useful error bounds. In Figure 5.3 (left) the relative errors on adaptive grids according to the *weighted* estimate and the ZZ-indicator are depicted, demonstrating η_{weight} to be more economical. Figure 5.3 (right) shows the structure of grids produced on the basis of η_{weight}.

Cells	$J(u_h)$	E^{rel}	Ratio		Cells	$J(u_h)$	E^{rel}	Ratio
256	2.378-3	1.048-3	24.45		580	9.953-4	1.062-1	1.87
1024	2.365-3	4.603-3	2.02		2320	1.056-3	5.112-2	1.38
2068	2.367-3	3.427-3	1.41		4480	1.084-3	2.598-2	2.14
4336	2.370-3	2.417-3	1.62		8860	1.097-3	1.452-2	2.74
9268	2.373-3	1.170-3	2.42		35440	1.103-3	9.046-3	1.92
19492	2.373-3	8.573-4	2.39		65596	1.106-3	6.222-3	2.37
40984	2.374-3	8.240-4	1.71		120712	1.108-3	4.826-3	2.71

Table 5.1: Numerical results for the first (left) and second (right) test-example: functional value $J(u_h)$, relative error E^{rel} and over-estimation factor Ratio.

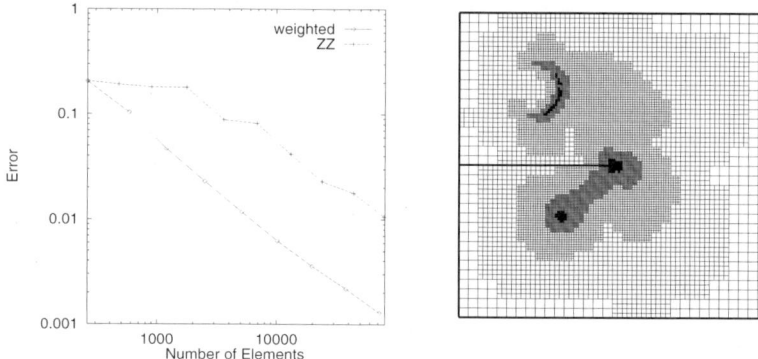

Figure 5.3: Relative error for the second example on adaptive grids according to the *weighted* estimate and the ZZ-indicator (left) demonstrating η_{weight} to be more economical. Structure of grids produced on the basis of η_{weight} (right).

Chapter 6

Signorini's problem

In this chapter, we consider Signorini's problem, a fundamental model situation for contact problems in elasticity, which is the basis for applications in the field of highspeed machining presented below. The corresponding classical notation reads (cf. Kikuchi and Oden [43])

$$-\operatorname{div}\sigma = f, \quad A\sigma = \varepsilon(u) \qquad \text{in } \Omega,$$
$$u = 0 \text{ on } \Gamma_D, \quad \sigma \cdot n = t \text{ on } \Gamma_N, \tag{6.1}$$
$$\left.\begin{array}{l} \sigma_T = 0, \quad (u_n - g)\sigma_n = 0 \\ u_n - g \leq 0, \quad \sigma_n \leq 0 \end{array}\right\} \text{ on } \Gamma_C.$$

This idealised model describes the deformation of an elastic body occupying the domain $\Omega \subset \mathbb{R}^3$, which is unilaterally supported by a frictionless rigid foundation. The displacement u and the corresponding stress tensor σ are caused by a body force f and a surface traction t along Γ_N. Along the portion Γ_D of the boundary the body is fixed and $\Gamma_C \subset \partial\Omega$ denotes the part which is a candidate contact surface. We use the notation $u_n = u \cdot n$, $\sigma_n = \sigma_{ij} n_i n_j$ and $\sigma_T = \sigma \cdot n - \sigma_n n$, where n is the outward normal of $\partial\Omega$, and g denotes the gap between Γ_C and the foundation.

Further, the deformation is supposed to be small, so that the strain tensor can be written as $\varepsilon(u) = \frac{1}{2}(\nabla u + \nabla u^T)$. The compliance tensor A is assumed to be symmetric and positive definite.

Similar to the previous chapter, we intend to derive efficient *a posteriori* error control techniques for this equation with special emphasis on local error phenomena, e.g., the error for stresses in the contact zone. In order to demon-strate the concept for our method for *a posteriori* error estimation, we first consider the simplified case

$$-\Delta u = f \ \text{in} \ \Omega \subset \mathbb{R}^2,$$
$$u = 0 \ \text{on} \ \Gamma_D,$$
$$u \geq 0, \quad \partial_n u \geq 0, \quad u \, \partial_n u = 0 \quad \text{on} \ \Gamma_C, \tag{6.2}$$

where $\Gamma_C = \partial\Omega \setminus \Gamma_D$ and $\partial_n u = \nabla u \cdot n$.

Problem (6.2) is to be solved by the finite element Galerkin method on adap-tively optimized meshes.

The basis for applying the finite element method to (6.2) is the formulation as a variational inequality, i.e., a solution $u \in K$ is sought fulfiling

$$(\nabla u, \nabla(\varphi - u)) \geq (f, \varphi - u) \qquad \forall \varphi \in K, \tag{6.3}$$

where we set $V = \{v \in H^1 \mid v = 0 \text{ on } \Gamma_D\}$ and $K = \{v \in V \mid v \geq 0 \text{ on } \Gamma_C\}$.

Eventually, the finite element approximation u_h of u in (6.3) is determined by

$$(\nabla u_h, \nabla(\varphi - u_h)) \geq (f, \varphi - u_h) \qquad \forall \varphi \in K_h, \tag{6.4}$$

with $K_h = V_h \cap K$. This finite dimensional problem can be shown to be uniquely solvable following the same line of arguments as in the continuous case. Optimal order *a priori* error estimates in the energy norm have been given for example in Falk [30] and Brezzi et al. [19]. Dobrowolski and Staib [23] show $\mathcal{O}(h)$-convergence in the energy norm without additional assumptions on the structure of the free boundary. Error estimates with respect to the L^∞-norm have been obtained, e.g., by Nitsche [52] based on a discrete maximum principle.

6.1   A posteriori error bounds

In order to derive an *a posteriori* error estimate for the scheme (6.4) we con-sider, following the strategies described in the previous chapter, the dual so-lution $z \in G$ of

$$(\nabla(\varphi - z), \nabla z) \geq J(\varphi - z) \qquad \forall \varphi \in G, \tag{6.5}$$

where $G = \{v \in V \mid v \geq 0 \text{ on } B_h \text{ and } \int_{\Gamma_C} \partial_n u(v + u_h) \leq 0\}$ and $B_h = \{x \in \Gamma_C \mid u_h(x) = 0\}$.

In order to show that $z + u - u_h \in G$, we observe $z + u - u_h \geq 0$ on B_h, since on $B_h \subset \Gamma_C$ we have $u \geq 0, u_h = 0$. Furthermore, there holds

$$\int_{\Gamma_C} \partial_n u((z + u - u_h) + u_h) = \int_{\Gamma_C} \partial_n u z \leq -\int_{\Gamma_C} \partial_n u u_h \leq 0.$$

Now, we can choose $\varphi = z + u - u_h$ as a testfunction in (6.5) and obtain

$$J(e) \leq (\nabla(u - u_h), \nabla z).$$

It is easily seen that $u_h \in W_h = \{v \in V \mid v \geq 0 \text{ on } B_h\} \cap V_h$, i.e., u_h coincides with the solution $\tilde{u}_h \in W_h$ of the discrete variational inequality

$$(f, \varphi - \tilde{u}_h) - (\nabla \tilde{u}_h, \nabla(\varphi - \tilde{u}_h)) \leq 0 \qquad \forall \varphi \in W_h. \tag{6.6}$$

With $z_h \in W_h$ and choosing $\varphi = u_h + z_h$ in (6.6) the first term on the right hand side of the identity

$$(\nabla(u - u_h), \nabla z_h) = \left\{(f, z_h) - (\nabla u_h, \nabla z_h)\right\} + (\nabla u, \nabla(z_h + u_h - u))$$
$$- (f, z_h + u_h - u) + \left\{(\nabla u, \nabla(u - u_h)) - (f, u - u_h)\right\} \quad \forall z_h \in W_h, \tag{6.7}$$

is negative. So is the last with $\varphi = u_h$ in (6.3). To sum up, we have shown the inequality

$$(\nabla(u - u_h), \nabla z_h) \leq (\nabla u, \nabla(z_h + u_h - u)) - (f, z_h + u_h - u), \tag{6.8}$$

for arbitrary $z_h \in W_h$.

Now, we proceed with estimating $J(e)$ by

$$J(e) \leq (\nabla(u - u_h), \nabla(z - z_h)) + (\nabla(u - u_h), \nabla z_h)$$
$$\leq (\nabla(u - u_h), \nabla(z - z_h)) + (\nabla u, \nabla(z_h + u_h - u)) - (f, z_h + u_h - u)$$
$$= (\nabla(u - u_h), \nabla(z - z_h)) + (\nabla u, \nabla(z + u_h - u)) - (f, z + u_h - u)$$
$$+ (\nabla u, \nabla(z_h - z)) - (f, z_h - z).$$

Due to $u\partial_n u = 0$ on Γ_C, we have with $z \in G$

$$(\nabla u, \nabla(z + u_h - u)) - (f, z + u_h - u) = \int_{\Gamma_C} \partial_n u(z + u_h) \leq 0.$$

Eventually, we obtain the *a posteriori* error estimate

$$J(e) \leq (f, z - z_h) - (\nabla u_h, \nabla(z - z_h)).$$ (6.9)

With standard techniques this can be exploited as follows. Elementwise integration by parts yields

$$J(e) \leq \sum_{T \in \mathbb{T}_h} \left\{ (f + \Delta u_h, z - z_h)_T - \tfrac{1}{2}([\partial_n u_h], z - z_h)_{\partial T} \right\},$$ (6.10)

where for interior interelement boundaries $[\partial_n u_h]$ denotes the jump of the normal derivative $\partial_n u_h$. Furthermore we set $[\partial_n u_h] = 0$ and $[\partial_n u_h] = \partial_n u_h$ on edges belonging to Γ_D and Γ_C respectively.

From (6.10), we deduce the *a posteriori* error bound

$$|J(e)| \leq \sum_{T \in \mathbb{T}_h} \omega_T \rho_T =: \eta_{weight},$$ (6.11)

with the local *residuals* ρ_T and *weights* ω_T defined by

$$\rho_T := h_T \| f + \Delta u_h \|_T + \tfrac{1}{2} h_T^{1/2} \| n \cdot [\nabla u_h] \|_{\partial T},$$
$$\omega_T := \max \left\{ h_T^{-1} \| z - z_h \|_T, h_T^{-1/2} \| z - z_h \|_{\partial T} \right\}.$$

Again, the weights ω_T cannot be determined analytically, but have to be computed by solving the dual problem numerically on the available mesh, yielding the approximation $\tilde{z}_h$. Now, interpreting z_h as a suitable interpolant of z, one uses the interpolation estimate

$$\omega_T \leq C_{i,T} h_T \| \nabla^2 z \|_T,$$ (6.12)

for $z \in H^2(T)$. For less regular z an estimate similar to (6.12) could be used involving a lower power of a local mesh size, which typically corresponds to higher values of ω_T. To evaluate the right hand side in (6.12) one may simply take second order difference quotients of the approximate dual solution $\tilde{z}_h$,

$$\omega_T \approx \tilde{\omega}_T := \tilde{C}_{i,T} h_T^2 | \nabla_h^2 \tilde{z}_h(x_T) |,$$ (6.13)

where x_T is the midpoint of element T. This results in approximate *a posteriori* error bounds using

$$\eta_{weight} \approx \sum_{T \in \mathbb{T}_h} \tilde{\omega}_T \rho_T.$$ (6.14)

It has been demonstrated in Becker and Rannacher [8] that this approximation has only minor effects on the quality of the resulting meshes. The *interpolation constant* C_i may be set equal to one for mesh designing.

Remark: The choice of (6.5) is not uniquely determined. Other approaches in *a priori* analysis in similar situations can be found, e.g., in Mosco [50]. Here, separate dual problems for the negative and positive part of the error are considered, but it seems to be difficult to exploit these techniques for *a posteriori* analysis, since data of the problem do not enter the estimate directly.

Remark: A posteriori error bounds for Signorini's problem

The weak solution $u \in K$ of (6.1) is defined by the variational formulation

$$a(u, \varphi - u) \geq F(\varphi - u) \qquad \forall \varphi \in K , \tag{6.15}$$

with the definitions

$$V = \{v \in H^1 \times H^1 \,|\, v = 0 \text{ on } \Gamma_D\}, \quad K = \{v \in V \,|\, v_n - g \leq 0\} ,$$

$$a(v, \varphi) = \int_\Omega A^{-1} \varepsilon(v) \varepsilon(\varphi) \quad \forall v, \varphi \in V , \qquad F(\varphi) = \int_\Omega f\varphi + \int_{\Gamma_N} t\varphi \quad \forall \varphi \in V .$$

As above, the discrete solution $u_h \in K_h = K \cap V_h \subset V$ is determined by

$$a(u_h, \varphi - u_h) \geq F(\varphi - u_h) \qquad \forall \varphi \in K_h . \tag{6.16}$$

Again, for estimating measures defined by $J(\cdot)$ of $e = u - u_h$, we employ $z \in G$, given by ·

$$a(\varphi - z, z) \geq J(\varphi - z) \qquad \forall \varphi \in G, \tag{6.17}$$

where $G = \{v \in V \,|\, v \geq 0 \text{ on } B_h \text{ and } F(v + u_h - u) - a(u, v + u_h - u) \geq 0\}$ and $B_h = \{x \in \Gamma_C \,|\, u_h(x) \cdot n = g(x)\}$. Eventually, the techniques used for the model case yield an *a posteriori* error estimate of the form (5.10) $|J(e)| \leq \sum_{T \in \mathbb{T}_h} \omega_T \rho_T$ with

$$\rho_T := h_T \|f + \text{div}(A^{-1}\varepsilon(u_h))\|_T + \tfrac{1}{2}h_T^{1/2}\|[n \cdot A^{-1}\varepsilon(u_h)]\|_{\partial T} ,$$

$$\omega_T := \max\left\{h_T^{-1}\|z - z_h\|_T, h_T^{-1/2}\|z - z_h\|_{\partial T}\right\} .$$

The approximation of the dual problem (6.17) may be realised as follows. Assuming B_h to be an appropriate approximation of B suggests to replace G by $\tilde{G} = \{v \in V \,|\, v = 0 \text{ on } B_h\}$ and solve a *linear elasticity problem* with Dirichlet boundary conditions on $\Gamma_D + B_h$.

6.2   Numerical results

Remark: The approximation of the dual problem (6.5)

$$(\nabla(\varphi - z), \nabla z) \geq J(\varphi - z) \qquad \forall \varphi \in G,$$

is realised as follows. Assuming $\partial_n u > 0$ on B_h and $\partial_n u = 0$ on $\Gamma_C \setminus B_h$ suggests to approximate G by $\tilde{G} = \{v \in V | v = 0 \text{ on } B_h\}$. Therefore, we only have to solve a *linear Dirichlet problem* with zero boundary conditions on $\Gamma_D + B_h$.

As a test example, we consider (6.2) on $\Omega = (0,1)^2$, $\Gamma_D = \{(x_1, x_2) \in \partial\Omega | x_1 = 0\}$ and right-hand side $f = 1000 \sin(2\pi x_1)$. The contact set $B = \{x \in \Gamma_C | u(x) = 0\}$ in this case is determined by $B = \{(x_1, x_2) \in \Gamma_C | x_1 \geq b\}$ with $b \approx 0.609374$ taken from u_{ref}. The structure of the solution is sketched in Figure 6.1 (left).

Applying an adaptive algorithm on the basis of the indicator η_{ZZ} yields locally refined grids with a structure shown in Figure 6.1, which can be compared with the grids based on η_{weight} for the following examples (Figures 6.2, 6.3 and 6.4).

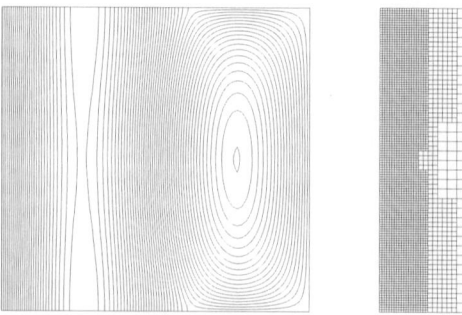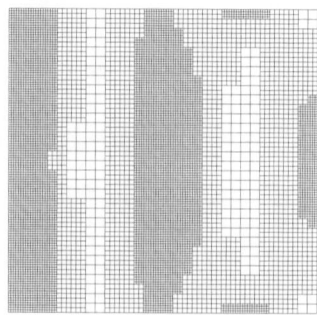

Figure 6.1: Isolines of the solution (left) and structure of grids produced on the basis of η_{ZZ} (right).

1) Point value: For the first test, we choose

$$J(\varphi) = \varphi(x_0), \qquad x_0 = (0.25, 0.25),$$

to control the point-error in x_0. The computational results are shown in Table 6.1. Evaluating *Ratio* shows the constant relation between *true* error and the corresponding estimation, and consequently it is demonstrated that the proposed approach to *a posteriori* error control gives useful error bounds. Figure 6.2 (left) demonstrates η_{weight} to produce a (monotonically) converging scheme with respect to the point value in contrast to the ZZ-approach. Figure 6.2 (right) shows the structure of grids produced on the basis of η_{weight}.

2) Mean value: As error functional for the second test, we choose

$$J(\varphi) = \int_B \partial_n \varphi,$$

to control the mean value of the normal derivative along the contact set B. In this case, the treatment of J, which is determined by derivatives of u, requires some additional care, since in this case the functional is *singular*, i.e. the dual solution is not properly defined on G. The remedy (cf. Becker and Rannacher [8] and see Rannacher and Suttmeier [55] for an application in linear elasticity) is to work with a regularised functional $J^\varepsilon(\cdot)$. In the present case, we set

$$J^\varepsilon(\varphi) = |B_\varepsilon|^{-1} \int_{B_\varepsilon} \partial_n \varphi \, dx,$$

where $B_\varepsilon := \{x \in \Omega, \, \text{dist}(x, B) < \varepsilon\}$. For each adaptive computation, the regularisation is done with the choice $\varepsilon = 0.5\eta_{weight}(u_h)$, where u_h is taken from the previous step.

The numerical results are presented in Table 6.2. Again, it is demonstrated, that the proposed approach to *a posteriori* error control gives useful error bounds. In Figure 6.3 (left) the relative errors on adaptive grids according to the *weighted* estimate and the ZZ-indicator are depicted, demonstrating η_{weight} to be most economical. Figure 6.3 (right) shows the structure of grids produced on the basis of η_{weight}.

3) Normal derivative: For the third test, we choose

$$J(\varphi) = \partial_n \varphi(x_0), \qquad x_0 = (1., 0.25),$$

to control the point error of the normal derivative in x_0. This example is chosen to indicate the applicability of the proposed techniques for our final goal of *a posteriori* error estimation of contact stresses in elasticity problems.

Again the treatment of J has to be done by regularisation analogously to the second example. Again the results presented in Table 6.3 and Figure 6.4 demonstrate η_{weight} to be reliable and efficient.

Cells	$J(u_h)$	E^{rel}	Ratio
928	2.928258e+01	1.446547e-03	2.07
1720	2.929928e+01	8.770673e-04	2.64
3148	2.930866e+01	5.572038e-04	2.97
5572	2.931476e+01	3.491901e-04	3.14
9604	2.931715e+01	2.676897e-04	3.40
16468	2.931918e+01	1.984655e-04	3.96
27724	2.932013e+01	1.660699e-04	2.99

Table 6.1: Numerical results for the first test-example: functional value $J(u_h)$, relative error E^{rel} and over-estimation factor Ratio.

Cells	$J(u_h)$	E^{rel}	Ratio
1840	-1.730673e+02	1.273645e-02	1.51
3256	-1.739847e+02	7.503137e-03	1.96
5980	-1.745723e+02	4.151169e-03	2.50
10528	-1.748522e+02	2.554478e-03	2.81
19204	-1.750084e+02	1.663434e-03	2.47
34540	-1.750833e+02	1.236167e-03	3.90
65212	-1.751289e+02	9.760411e-04	3.85
122284	-1.751526e+02	8.408443e-04	2.67

Table 6.2: Numerical results for the second test-example: functional value $J(u_h)$, relative error E^{rel} and over-estimation factor Ratio.

Cells	$J(u_h)$	E^{rel}	Ratio
628	1.146645e+02	2.396903e-03	1.59
1312	1.148638e+02	6.629546e-04	2.17
2548	1.149107e+02	2.549156e-04	2.17
4912	1.149301e+02	8.613189e-05	3.27
9208	1.149339e+02	5.307117e-05	2.29
17200	1.149363e+02	3.219071e-05	2.08
31468	1.149372e+02	2.436054e-05	1.71

Table 6.3: Numerical results for the third test-example: functional value $J(u_h)$, relative error E^{rel} and over-estimation factor Ratio.

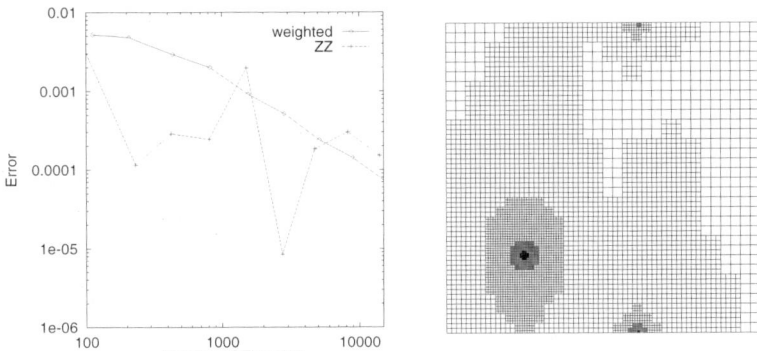

Figure 6.2: Relative error for the first example on adaptive grids according to the *weighted* estimate and the ZZ-indicator (left) demonstrating η_{weight} to produce a (monotonically) converging scheme with respect to the point value. Structure of grids produced on the basis of η_{weight} (right).

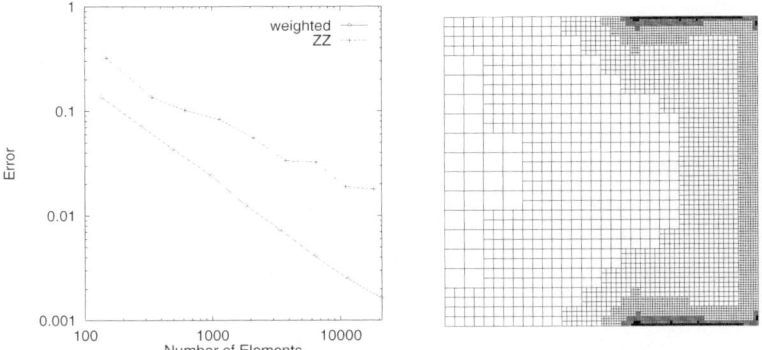

Figure 6.3: Relative error for the second example on adaptive grids according to the *weighted* estimate and the ZZ-indicator (left) demonstrating η_{weight} to be most economical. Structure of grids produced on the basis of η_{weight} (right).

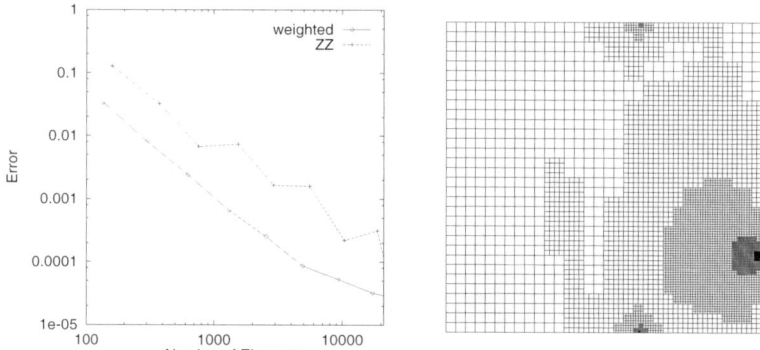

Figure 6.4: Relative error for the third example on adaptive grids according to the *weighted* estimate and the ZZ-indicator (left) demonstrating η_{weight} to be most economical. Structure of grids produced on the basis of η_{weight} (right).

6.3   A posteriori controlled boundary approximation

In some applications presented below, we have to deal with a domain with a non-polygonal boundary. A detailed analysis of the effect of boundary approximation for FE-schemes for variational equations is given by Dörfler and Rumpf in [24]. Referring to this paper for technical details, we ommit the case of Γ_D to be non-polygonal and focus on the case of Γ_C to be a convex and piecewise smooth part of the boundary.

Here, we sketch basic ideas for deriving *a posteriori* error bounds for FE-discretisations of variational inequalities on domains with curved boundaries for the simplified Signorini problem. Defining $a(.,.) = (\nabla., \nabla.)$, the variational settings corresponding to (6.2) determin $u \in K$ and $u_h \in K_h$ by

$$a(u, \varphi - u)_\Omega \geq (f, \varphi - u)_\Omega \qquad \forall \varphi \in K, \tag{6.18}$$

$$a(u_h, \varphi - u_h)_{\Omega_h} \geq (f, \varphi - u_h)_{\Omega_h} \qquad \forall \varphi \in K_h, \tag{6.19}$$

with the definitions

$$K = \{v \in H^1(\Omega) \mid v = 0 \text{ on } \Gamma_D \text{ and } v \geq 0 \text{ on } \Gamma_C\},$$
$$K_h = \{v \in C^0(\Omega_h) \mid v_{|T} \text{ is linear on } T \in \mathbb{T}_h, v(p) = 0 \text{ for a vertex } p \in \Gamma_D,$$
$$v(p) \geq 0 \text{ for a vertex } p \in \Gamma_C\}.$$

Here, $\mathbb{T}_h := \Omega_h$ is a union of triangles T. Vertices on $\partial\Omega_h$ are assumed to lie on $\partial\Omega$. To each element $T \in \mathbb{T}_h$ with $\partial T \cap \partial\Omega \neq \emptyset$ there naturally belongs a section S_T of $\Omega \setminus \Omega_h$, such that $\mathbb{T}_h^C := \Omega \setminus \Omega_h = \bigcup_{T \in \mathbb{T}_h} S_T$.

In the following $\tilde{u}_h$ denotes the linear extension of u_h to Ω and δ is chosen, such that $\delta = 0$ on Ω_h and $\tilde{u}_h + \delta \geq 0$ on Γ_C. After these preparations we start the estimation of $\|\nabla e\|$.

$$
\begin{aligned}
\|\nabla e\|_\Omega^2 &= a(u - \tilde{u}_h, e)_\Omega \\
&= -a(\tilde{u}_h, e)_{\Omega \setminus \Omega_h} + a(u, e)_\Omega - a(u_h, e)_{\Omega_h} \\
&= -a(\tilde{u}_h, e)_{\Omega \setminus \Omega_h} + a(u, e)_\Omega - a(u_h, e - I_h e)_{\Omega_h} \\
&\quad \underbrace{-a(u_h, I_h e)_{\Omega_h} + (f, I_h e)_{\Omega_h}}_{I} - (f, I_h e)_{\Omega_h} \\
&\leq -a(\tilde{u}_h, e)_{\Omega \setminus \Omega_h} \underbrace{+a(u, u - \tilde{u}_h)_\Omega - (f, u - \tilde{u}_h)_\Omega}_{II} \\
&\quad + (f, e)_\Omega - a(u_h, e - I_h e)_{\Omega_h} - (f, I_h e)_{\Omega_h} \\
&\leq -a(\tilde{u}_h, e)_{\Omega \setminus \Omega_h} + a(u, \delta)_{\Omega \setminus \Omega_h} - (f, \delta)_{\Omega \setminus \Omega_h} \\
&\quad + (f, e)_\Omega - a(u_h, e - I_h e)_{\Omega_h} - (f, I_h e)_{\Omega_h}.
\end{aligned}
$$

The terms I and II in this calculation are estimated by

$$I \leq 0,$$
$$II \leq a(u, \delta)_\Omega - (f, \delta)_\Omega = a(u, \delta)_{\Omega \setminus \Omega_h} - (f, \delta)_{\Omega \setminus \Omega_h},$$

recalling, for I, that there holds $I_h e = I_h u - u_h$ and $I_h u$ is an admissible testfunction in the discrete setting (6.19). The second estimate is obtained by testing the continuous formulation (6.18) with $\tilde{u}_h + \delta$.

We summarise the above yielding

$$a(u - \tilde{u}_h, e)_\Omega \leq \underbrace{(f, e - I_h e)_{\Omega_h} - a(u_h, e - I_h e)_{\Omega_h}}_{i}$$

$$+ \underbrace{(f, e)_{\Omega \setminus \Omega_h}}_{ii} - a(\tilde{u}_h, e)_{\Omega \setminus \Omega_h} + \underbrace{a(u, \delta)_{\Omega \setminus \Omega_h} - (f, \delta)_{\Omega \setminus \Omega_h}}_{iii} \,. \quad (6.20)$$

Term i is evaluated similar to the previous section. The treatment of term ii employs $\|e\|_{\Omega \setminus \Omega_h} \leq c \|\nabla e\|_\Omega$, eventually giving

$$\tfrac{1}{2} \|\nabla e\|_\Omega^2 \leq a(u, \delta)_{\Omega \setminus \Omega_h} - (f, \delta)_{\Omega \setminus \Omega_h} + c \sum_{T \in \mathbb{T}_h} \left(h_T^2 \|f + \Delta \tilde{u}_h\|_T^2 + \tfrac{1}{2} h_T \|n \cdot [\nabla \tilde{u}_h]\|_{\partial T}^2 \right)$$

$$+ c \sum_{S_T \in \mathbb{T}_h^C} \left(\|f\|_{S_T}^2 + \|\nabla \tilde{u}_h\|_{S_T}^2 \right) \,.$$

For practical evaluation of this error bound, we propose to replace u in the first term on the right-hand side simply by $\tilde{u}_h$. Future work concerning the subject of this section has to deal with Γ_C to be non-convex.

Chapter 7

Strang's problem

As a third example, we discuss the situation from elasto-plasticity theorie, already presented in Chapter 3. In this case we have to deal with a system of unknowns yielding a mixed variational setting. Furthermore, in contrast to the previous examples, the imposed inequality constraints are nonlinear.

We recall, assuming $\gamma = 0$ for the hardening parameter, that the mathematical model seeks for a scalar displacement u in the vertical direction and a stress vector $\sigma = (\sigma_1, \sigma_2)$ as functions on Ω. The plastic behaviour of the material is taken into account by the nonlinear restriction $|\sigma| \leq 1$. This results in the system

$$- \operatorname{div} \sigma = f, \quad \sigma = \Pi \nabla u \quad \text{in } \Omega, \tag{7.1}$$
$$u = 0 \quad \text{on } \partial\Omega,$$

where Π denotes the pointwise projection onto the circle with radius 1.

In order to give a weak form for (7.1), we set

$$L^2(\Omega)^2 := L^2(\Omega, \mathbb{R}^2),$$
$$\Pi H := \left\{ \tau \in L^2(\Omega)^2, |\tau| - 1 \leq 0 \right\},$$
$$V := \left\{ u \in BD(\Omega), \ u = 0 \text{ on } \partial\Omega \right\},$$

where $BD(\Omega) = \{ v \in L^2(\Omega) \mid \int |\nabla v| < \infty \}$ and $|\tau|^2 = \tau_1^2 + \tau_2^2$. Now, similar to the approach in Johnson [40], the solution $\{\sigma, u\} \in \Pi H \times V$ is determined by the variational inequality

$$(\sigma, \tau - \sigma) - (\nabla u, \tau - \sigma) + (\sigma, \nabla \varphi) \geq (f, \varphi) \qquad \forall \{\tau, \varphi\} \in \Pi H \times V. \quad (7.2)$$

Existence of the solution and uniquenes for the stresses σ have been proven, e.g., by Johnson [40].

The finite dimensional space $\Pi H_h \times V_h$ for a stable discretisation on triangulations only consisting of triangular elements for this saddle point problem is determined by approximating each component of the stresses with elementwise constant functions for ΠH_h. V_h is constructed by the standard linear shape functions. *A priori* error analysis for this discretisation can be found, e.g., in Falk and Mercier [31].

7.1 A posteriori error bounds

In order to introduce a compact variational setting and its corresponding discretisation in the forms (1.4) and (1.5),

$$A(U, \varphi - U) \geq F(\varphi - U) \qquad \forall \varphi \in \mathbf{K},$$
$$A(U_h, \varphi - U_h) \geq F(\varphi - U_h) \qquad \forall \varphi \in \mathbf{K}_h,$$

we identify U, U_h with the pairs $\{\sigma, u\}, \{\sigma_h, u_h\}$. The weak formulation for the example under consideration is determined by the definitions

$$A(\{\sigma, u\}, \{\tau, \varphi\}) = (\sigma, \tau) - (\nabla u, \tau) + (\sigma, \nabla \varphi),$$
$$F(\{\tau, \varphi\}) = f(\varphi),$$
$$\mathbf{V} = \{\{\tau, \varphi\} \in L^2(\Omega)^2 \times BD(\Omega)\}, \qquad \mathbf{K} = \{\{\tau, \varphi\} \in \Pi H \times BD(\Omega)\}.$$

Oriented at the obstacle problem, we consider the dual problem

$$J(\tau - \pi, \varphi) \leq (\pi, \tau - \pi) - (\nabla \varphi, \pi) + (\tau - \pi, \nabla z) \qquad \forall \{\tau, \varphi\} \in G, \quad (7.3)$$

with dual solution $Z = \{\pi, z\} \in G$ and in a first run $G = \{ v \in \mathbf{V} \mid F(v + U_h - U) - A(U, v + U_h - U) \geq 0 \}$. Choosing $\{\pi + (\sigma - \sigma_h), u - u_h\}$ as a test function, we obtain

$$
\begin{aligned}
J(U - U_h) &\leq A(U - U_h, Z) \\
&= A(U - U_h, Z - Z_h) + A(U - U_h, Z_h) \\
&\leq A(U - U_h, Z - Z_h) + A(U, Z_h + U_h - U) \\
&\quad - F(Z_h + U_h - U) + (\nabla u_h - \sigma_h, \pi_h) \\
&\leq A(U - U_h, Z - Z_h) + \underbrace{A(U, Z + U_h - U) - F(Z + U_h - U)}_{\leq 0} \\
&\quad + A(U, Z_h - Z) - F(Z_h - Z) + (\nabla u_h - \sigma_h, \pi_h) \, .
\end{aligned}
\tag{7.4}
$$

To get the second inequality (7.4), we use the estimate

$$
\begin{aligned}
A(U - U_h, Z_h) &= (\sigma - \sigma_h, \pi_h) - (\nabla(u - u_h), \pi_h) + (\sigma - \sigma_h, \nabla z_h) \\
&= f(z_h) - (\sigma_h, \pi_h) + (\nabla u_h, \pi_h) - (\sigma_h, \nabla z_h) \\
&\quad + (\sigma, \pi_h + \sigma_h - \sigma) - (\nabla u, \pi_h + \sigma_h - \sigma) \\
&\quad + (\sigma, \nabla(z_h + u_h - u)) - f(z_h + u_h - u) \\
&\quad + \underbrace{(\sigma, \sigma - \sigma_h) - (\nabla u, \sigma - \sigma_h)}_{\leq 0} + \underbrace{(\sigma, \nabla(u - u_h)) - f(u - u_h)}_{=0} \\
&\leq A(U, Z_h + U_h - U) - F(Z_h + U_h - U) + (\nabla u_h - \sigma_h, \pi_h) \, .
\end{aligned}
$$

Inserting our definitions, eventually we obtain

$$
J(\{\sigma - \sigma_h, u - u_h\}) \leq (\nabla u_h - \sigma_h, \pi) + f(z - z_h) - (\sigma_h, \nabla(z - z_h)) \, .
$$

This suggests an improved choice of G:

$$
\begin{aligned}
G = \{\{\tau, \varphi\} \in \mathbf{V} \mid (\nabla u_h - \sigma_h, \tau) \leq 0\} \\
\cap \{v \in \mathbf{V} \mid F(v + U_h - U) - A(U, v + U_h - U) \geq 0\} \, .
\end{aligned}
$$

Now, it remains to show $\theta = \{\pi + (\sigma - \sigma_h), (u - u_h)\}$ to be an admissible testfunction for the dual problem (7.3), i.e., $\{\pi, \varphi\} \in G$ implies $\theta \in G$.

Proof.
i) With σ_i denoting the L^2-Projektion, term II in

$$
\begin{aligned}
(\nabla u_h - \sigma_h, \pi + \sigma - \sigma_h) &= (\nabla u_h - \sigma_h, \pi + \sigma - \sigma_i + \sigma_i - \sigma_h) \\
&= \underbrace{(\nabla u_h - \sigma_h, \pi)}_{I} + \underbrace{(\nabla u_h - \sigma_h, \sigma - \sigma_i)}_{II} + \underbrace{(\nabla u_h - \sigma_h, \sigma_i - \sigma_h)}_{III} \leq 0 \, .
\end{aligned}
$$

is zero. For $\{\pi, \varphi\} \in G$ term I is non-positive. Furthermore σ_i is an admissible test function in the discrete variational formulation

$$(\sigma_h - \nabla u_h, \tau - \sigma_h) \geq 0 \qquad \forall \tau \in \Pi H_h \,,$$

which implies the term III to be negative or zero. Thus θ fulfils the first condition in the definition of G.

ii) Choosing $\varphi = U_h$ as a testfunction in the continuous variational inequality (1.4) and because $Z \in G$ there holds

$$0 \leq F(U - U_h) - A(U, U - U_h) + F(Z + U_h - U) - A(U, Z + U_h - U)$$
$$= F((Z + U - U_h) + U_h - U) - A(U, (Z + U - U_h) + U_h - U) \,.$$

Thus θ fulfils the second condition in the definition of G. $\qquad\square$

Summarising, our *a posteriori* error estimate can be written in the form

$$J(\{\sigma - \sigma_h, u - u_h\}) \leq \sum_{T \in \mathbb{T}_h} \omega_T \rho_T \,,$$

with the local *residuals* ρ_T and *weights* ω_T defined by

$$\rho_T = \max \left\{ h_T \|f\|_T, \tfrac{1}{2} h^{1/2} \|n \cdot [\sigma_h]\|_{\partial T} \right\} \,,$$
$$\omega_T = \max \left\{ h_T^{-1} \|z - z_h\|_T, h_T^{-1/2} \|z - z_h\|_{\partial T} \right\} \,,$$

where $[.]$ denotes the jumps of σ_h across the interelement boundary.

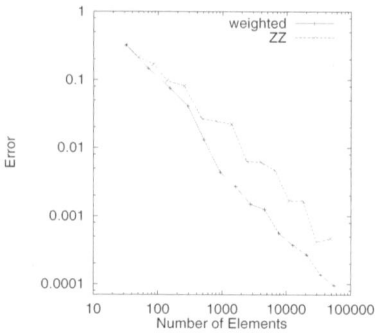 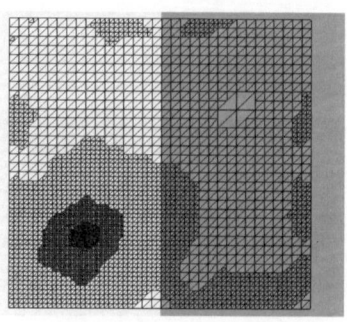

Figure 7.1: Relative errors for Strang's example on adaptive grids according to the *weighted* estimate and the ZZ-indicator (left) demonstrating η_{weight} to be more economical. Structure of grids produced on the basis of η_{weight} (right).

7.2 Numerical results

As a test example for Strang's problem, we consider (7.1) on $\Omega = (0,1)^2$ with $f = 2\sin(\pi x)\sin(\pi y)$. The value for the yield condtion is 0.3. For this test, we choose

$$J(\{\tau, \varphi\}) = \tau_1(x_0), \qquad x_0 = (0.25, 0.25),$$

to control the point-error for the stress value in x_0.

Results: The numerical results are presented in Tables 7.1. Evaluating *Ratio* shows the constant relation between *true* error and the corresponding estimation, and consequently it is demonstrated that the proposed approach to *a posteriori* error control gives useful error bounds.

In Figure 7.1 the relative errors on adaptive grids according to the *weighted* estimate and the ZZ-indicator are depicted, demonstrating η_{weight} to be more economical. Furthermore the structure of grids produced on the basis of η_{weight} is shown.

Cells	$J(u_h)$	E^{rel}	Ratio
1610	1.655007e-01	2.707442e-03	2.03
2726	1.657008e-01	1.501657e-03	2.37
4544	1.657422e-01	1.252184e-03	1.65
7607	1.658569e-01	5.610124e-04	2.44
12623	1.658875e-01	3.766195e-04	2.25
20717	1.659049e-01	2.717686e-04	1.86
34064	1.659273e-01	1.367882e-04	2.43
55436	1.659342e-01	9.520940e-05	2.10

Table 7.1: Numerical results for Strang's problem: functional value $J(u_h)$, relative error E^{rel} and over-estimation factor Ratio.

Chapter 8

General concept

In this chapter, collecting the experiences from the three examples treated above, we propose a framework for deriving *a posteriori* error estimates for the scheme (1.5), where the error $e = U - U_h$ is measured in terms of a linear functional $J(\cdot)$ defined on $\mathbf{V}$ or a suitable subspace. The main ingredients are a generalisation of the Galerkin orthogonality relation in the context of variational equations and a suitable duality argument.

8.1   Orthogonality relation

A crucial property for the error analysis of the nonrestricted problem and its FE-discretisation with solutions $\mathcal{U}$ and $\mathcal{U}_h$ respectively

$$\mathcal{U} \in \mathbf{V} : \quad A(\mathcal{U}, \varphi) = F(\varphi) \qquad \forall \varphi \in \mathbf{V},$$
$$\mathcal{U}_h \in \mathbf{V}_h : \quad A(\mathcal{U}_h, \varphi) = F(\varphi) \qquad \forall \varphi \in \mathbf{V}_h,$$

is the Galerkin orthogonality relation

$$A(\mathcal{U} - \mathcal{U}_h, Z_h) = 0 \qquad \forall Z_h \in \mathbf{V}_h. \tag{8.1}$$

Now, we formulate a generalisation of this equality for the case of variational inequalities in the form

$$U \in \mathbf{V} : \quad F(\varphi - U) - A(U, \varphi - U) \leq 0 \qquad \forall \varphi \in \mathbf{K}, \tag{8.2}$$
$$U_h \in \mathbf{V}_h : \quad F(\varphi - U_h) - A(U_h, \varphi - U_h) \leq 0 \qquad \forall \varphi \in \mathbf{K}_h. \tag{8.3}$$

After these preparations, we generalise the orthogonality relation (8.1) to the Galerkin estimate

Lemma 8.1.1. *Let $W_h^0 \subset \{Z_h \in \mathbf{V}_h \mid F(Z_h) - A(U_h, Z_h) \leq 0\}$, then there holds the estimate*

$$A(U - U_h, Z_h) \leq A(U, Z_h + U_h - U) - F(Z_h + U_h - U) \quad \forall Z_h \in W_h^0. \quad (8.4)$$

Proof. With $Z_h \in W_h^0$ the first term on the right-hand side of the identity

$$A(U - U_h, Z_h) = (F(Z_h) - A(U_h, Z_h)) + A(U, Z_h + U_h - U) - F(Z_h + U_h - U)$$
$$+ (A(U, U - U_h) - F(U - U_h)) \quad \forall Z_h \in W_h^0 \quad (8.5)$$

is negative. So is the last with $\varphi = U_h$ in (8.2), which completes the proof. $\square$

The following lemma describes the construction of a W_h^0, which is useful for applications with structures comparable to the obstacle problem.

Lemma 8.1.2. *Let W_h, with $K_h \subset W_h \subset V_h$, be constructed, such that the discrete solution of (8.3) is equivalently characterised by*

$$F(\varphi - U_h) - A(U_h, \varphi - U_h) \leq 0 \qquad \forall \varphi \in W_h. \quad (8.6)$$

Furthermore let $W_h^0 = \{Z_h \in V_h \mid U_h + Z_h \in W_h\}$. Then there holds the estimate

$$F(Z_h) - A(U_h, Z_h) \leq 0 \qquad \forall Z_h \in W_h^0. \quad (8.7)$$

Proof. With $Z_h \in W_h^0$ and choosing $\varphi = U_h + Z_h$ in (8.6) ensures the result of the lemma. $\square$

8.2   Duality argument

The next step in our approach to *a posteriori* error control for $J(e)$ is the dual problem

$$Z \in G: \quad A(\varphi - Z, Z) \geq J(\varphi - Z) \qquad \forall \varphi \in G, \quad (8.8)$$

with solution $Z \in G$, where the set G is assumed to fulfil the properties

$$Z \in G \Rightarrow Z + U - U_h \in G, \tag{8.9}$$

$$G \subset \{ v \in \mathbf{V} \mid F(v + U_h - U) - A(U, v + U_h - U) \geq 0 \}. \tag{8.10}$$

Remark: $Z + U - U_h$ fulfils the condition in (8.10) by construction, since with $\varphi = U_h$ in (8.2) and because $Z \in G$ there holds

$$0 \leq F(U - U_h) - A(U, U - U_h) + F(Z + U_h - U) - A(U, Z + U_h - U)$$
$$= F((Z + U - U_h) + U_h - U) - A(U, (Z + U - U_h) + U_h - U).$$

Estimation:

Now, choosing $\varphi = Z + U - U_h$ as a testfunction in (8.8) and employing the orthogonality estimate (8.4), we get for $Z_h \in W_h^0$

$$\begin{aligned}
J(e) &\leq A(U - U_h, Z) \\
&\leq A(U - U_h, Z - Z_h) + A(U - U_h, Z_h) \\
&\leq A(U - U_h, Z - Z_h) + A(U, Z_h + U_h - U) - F(Z_h + U_h - U) \\
&\leq A(U - U_h, Z - Z_h) + A(U, Z + U_h - U) - F(Z + U_h - U) \\
&\quad + A(U, Z_h - Z) - F(Z_h - Z).
\end{aligned}$$

We recall, see (8.10), that for $Z \in G$ there holds $A(U, Z + U_h - U) - F(Z + U_h - U) \leq 0$. Eventually, we obtain the following

Theorem 8.2.1. *For scheme* (8.3), *there holds the a posteriori error estimate*

$$|J(e)| \leq F(Z - Z_h) - A(U_h, Z - Z_h) \qquad \forall Z_h \in W_h^0. \tag{8.11}$$

8.3 Modification

Above, the error $U - U_h$ enters the definition of G. For semi-definite problems, i.e., $A(v, v) \geq 0$, we can alternatively proceed as follows.

The set G is assumed to fulfil the properties

$$Z \in G \Rightarrow Z + U - U_h \in G, \tag{8.12}$$

$$G \subset \{ v \in \mathbf{V} \mid F(v) - A(U, v) \geq 0 \}. \tag{8.13}$$

Remark: $Z + U - U_h$ fulfils the condition in (8.13) by construction, since with $\varphi = U_h$ in (8.2) and because $Z \in G$ there holds

$$0 \leq F(U - U_h) - A(U, U - U_h) + F(Z) - A(U, Z)$$
$$= F(Z + U - U_h) - A(U, Z + U - U_h).$$

Estimation:

Now, choosing $\varphi = Z + U - U_h$ as a testfunction in (8.8) and employing the orthogonality estimate (8.4), we get for $Z_h \in W_h^0$

$$
\begin{aligned}
J(e) &\leq A(U - U_h, Z) \\
&\leq A(U - U_h, Z - Z_h) + A(U - U_h, Z_h) \\
&\leq A(U - U_h, Z - Z_h) + A(U, Z_h + U_h - U) - F(Z_h + U_h - U) \\
&\leq A(U - U_h, Z - Z_h) + A(U, Z + U_h - U) - F(Z + U_h - U) \\
&\qquad + A(U, Z_h - Z) - F(Z_h - Z) \\
&= F(Z - Z_h) - A(U_h, Z - Z_h) \\
&\qquad + \underbrace{A(U, Z) - F(Z)}_{\leq 0} + A(U, U_h - U) - F(U_h - U) \\
&\leq F(Z - Z_h) - A(U_h, Z - Z_h) \\
&\qquad + \underbrace{A(U - U_h, U_h - U)}_{\leq 0} + A(U_h, U_h - U) - F(U_h - U) \\
&\leq F(Z - Z_h) - A(U_h, Z - Z_h) \\
&\qquad + A(U_h, U_h - U_i + U_i - U) - F(U_h - U_i + U_i - U) \\
&= F(Z - Z_h) - A(U_h, Z - Z_h) + F(e - e_i) - A(U_h, e - e_i),
\end{aligned}
$$

where the subscript i denotes a standard interpolation operator. Eventually, we obtain the following

Theorem 8.3.1. *For scheme* (8.3), *there holds the a priori/a posteriori error estimate*

$$|J(e)| \leq F(Z - Z_h) - A(U_h, Z - Z_h)$$
$$+ F(e - e_i) - A(U_h, e - e_i) \qquad \forall Z_h \in W_h^0. \quad (8.14)$$

Remark: As indicated above, in general, Z cannot be determined analytically, but has to be replaced by $\tilde{Z}$, where $\tilde{Z}$ is the solution of a (commonly discrete) dual problem approximating the original one.

The treatment of the dual problem (8.8) may be realised as follows. We recall the definitions of W_h^0 and the modified version of G, i.e.,

$$W_h^0 \subset \{Z_h \in \mathbf{V}_h \mid F(Z_h) - A(U_h, Z_h) \leq 0\} \tag{8.15}$$

$$G \subset \{\, v \in \mathbf{V} \mid F(v) - A(U, v) \geq 0 \,\}. \tag{8.16}$$

This suggests, after replacing the unknown solution U in (8.16) by U_h, to approximate G by $\tilde{G} = \{\, v \in \mathbf{V} \mid F(v) - A(U_h, v) = 0 \,\}$.

8.3.1 Example

In order to obtain a pure *a posteriori* estimate, one may treat the term $F(e - e_i) - A(U_h, e - e_i)$ by using interpolation and standard energy error estimates.

Applying the modified concept to the obstacle problem yields

Theorem 8.3.2. *For the scheme (5.3) there holds the a priori/a posteriori error bound*

$$|J(e)| \leq \sum_{T \in \mathbb{T}_h} \omega_T \rho_T + \sum_{T \in \mathbb{T}_h} \omega_T^e \rho_T , \tag{8.17}$$

with local residuals ρ_T and weights ω_T, ω_T^e defined by

$$\rho_T := h_T \|f + \Delta u_h\|_T + \tfrac{1}{2} h_T^{1/2} \|n \cdot [\nabla u_h]\|_{\partial T} ,$$

$$\omega_T := \max \left\{ h_T^{-1} \|z - z_h\|_T, h_T^{-1/2} \|z - z_h\|_{\partial T} \right\} ,$$

$$\omega_T^e := \max \left\{ h_T^{-1} \|e - e_i\|_T, h_T^{-1/2} \|e - e_i\|_{\partial T} \right\} ,$$

where for interior interelement boundaries $[\partial_n u_h]$ denotes the jump of the normal derivative $\partial_n u_h$.

Eventually, treating the second sum in (8.17) by using the interpolation estimate $\omega_T^e \leq C_{i,T} \|\nabla e\|_T$, and employing a standard *a posteriori* energy error estimate, we obtain

$$\sum_{T \in \mathbb{T}_h} \omega_T^e \rho_T \leq \sum_{T \in \mathbb{T}_h} C_{i,T}^2 \rho_T^2 + \sum_{T \in \mathbb{T}_h} \|\nabla e\|^2 \leq 2 \sum_{T \in \mathbb{T}_h} C_{i,T}^2 \rho_T^2 .$$

Chapter 9

Lagrangian formalism

In this chapter, we will discuss a further application of the theory of VI's, namely the torsion problem. It turns out to be convenient to treat this problem by a Lagrangian formalism. The approach offers several alternatives for the numerical analysis of variational inequalities. We mention the iterative solution process of the discrete problems and focus on new possibilities for *a priori* error analysis.

9.1 Torsion problem

Studying the numerical analysis of variational inequalities, one standard example is the torsion problem (cf. Glowinski [33]). The physical situation behind is a cylindrical tree consisting of isotropic elastic perfectly plastic material, bounded by two plane sections Γ_0, Γ_1. The flow behaviour follows the von Mises rule. The bar is subjected to the volume force $F = 0$ in Ω and to the boundary conditions

$$\sigma \cdot n = 0 \qquad \text{on } \Gamma_2 = \partial\Omega \setminus (\Gamma_0 \cup \Gamma_1)$$

$$\sigma_{33} = 0 \qquad \text{on } \Gamma_0, \Gamma_1$$

$$U_1 = U_2 = 0 \qquad \text{on } \Gamma_0$$

$$U_1 = -\alpha H x_2, \quad U_2 = \alpha H x_1, \qquad \text{on } \Gamma_1,$$

with given $\alpha > 0$ and H denotes the height of the bar.

Following Duvaut and Lions [26], the resulting stress field has the form

$$\sigma_{ij}(x) = \begin{cases} \sigma_{ij}(x_1, x_2) & \text{if } (ij = 13) \text{ or } (ij = 23) \\ 0 & \text{otherwise} \end{cases} .$$

Assuming a simply connected cross section one introduces a scalar potential u, which determines the stress field by

$$\sigma_{13} = \frac{\partial u}{\partial x_2}, \qquad \sigma_{23} = -\frac{\partial u}{\partial x_1} .$$

Defining

$$a(v, \varphi) = (\nabla v, \nabla \varphi), \quad f = 2\mu\alpha ,$$

where the constant $\mu > 0$ denotes the shear modulus, the unknown function u is obtained as minimiser of the functional

$$J(\varphi) = \frac{1}{2} a(\varphi, \varphi) - (f, \varphi) \tag{9.1}$$

over the convex set

$$K = \{\varphi \in H_0^1 \mid |\nabla\varphi| \le 1\} .$$

The variational setting in form of an inequality is given by

$$a(u, \varphi - u) \ge (f, \varphi - u) \qquad \forall \varphi \in K . \tag{9.2}$$

Problem (9.2) is uniquely solvable (cf. Glowinski [33]) and, under appropriate smoothness conditions on the boundary and data, the solution is known to satisfy the regularity result $u \in W^{2,p}(\Omega), 1 < p < \infty$ (see Brézis and Stampacchia [18]).

The solution $u \in K$ is approximated by $u_h \in K_h \subset K$ with

$$a(u_h, \varphi - u_h) \ge (f, \varphi - u_h) \qquad \forall \varphi \in K_h , \tag{9.3}$$

where

$$K_h = \{\varphi \in V_h \mid |\nabla\varphi| \le 1\}, \quad V_h = \{\varphi \in H_0^1(\Omega) \mid \varphi \text{ linear on } T \in \mathbb{T}_h\} .$$

In what follows, a priori error estimates in the energy norm are derived for the discretisation (9.3) of (9.2). In the next section, we recall the suboptimal results of a direct approach following Glowinski [33]. Subsequently, by means of duality, we are able to derive an almost optimal order of convergence: a suitable Lagrange functional, where the primal variable of the corresponding saddle point coincides with the solution of the original problem, is analysed, yielding improved error estimates.

9.2   A suboptimal error estimate

Following Glowinski [33], a first error estimate is obtained as indicated below. Using $u_h, \varphi_h \in K_h \subset K$ in (9.2),(9.3) and adding the two inequalities gives

$$a(u_h - u, u_h - u) \le a(\varphi_h - u, u_h - u) + a(u, \varphi_h - u) - (f, \varphi_h - u),$$

and thus,

$$\|\nabla(u_h - u)\|^2 \le \frac{1}{2}\|\nabla(u_h - u)\|^2 + \frac{1}{2}\|\nabla(\varphi_h - u)\|^2 + (\|\Delta u\| + \|f\|)\|\varphi_h - u\|.$$

Now the problem arises to construct a suitable $\varphi_h \in K_h$ with optimal approximation properties with respect to u. As described in Glowinski [33], the standard nodal interpolant usually does not fulfil $I_h u \in K$. Modifying $I_h u$ to $r_h u \in K$ by

$$r_h u = I_h u \cdot (1 + Ch^{1-2/p}\|u\|_{2,p})^{-1}$$

yields the suboptimal result

$$\|\nabla(r_h u - u)\| \le \|\nabla(I_h u - u)\| + C_1 h^{1-2/p}\|u\|_{2,p},$$

and even worse,

$$\|r_h u - u\| \le \|I_h u - u\| + C_2 h^{1-2/p}\|u\|_{2,p}.$$

In practice, where $f \in L^\infty$, the convergence behaviour of the discretisation error measured in the energy norm tends to the order $\mathcal{O}(h^{1/2})$.

Remark: Introducing

$$B = \{x \in \Omega \mid |\nabla u| = 1\},$$
$$B_h = \{T \in \mathbb{T}_h \mid |\nabla u_h| = 1\},$$
$$W_h = \{\varphi \in V_h \mid |\nabla\varphi| \le 1 \text{ on } B_h\},$$

we notice that

$$a(u, \varphi_h - u) = (f, \varphi_h - u) \qquad \text{on } \Omega \setminus B \,.$$

Furthermore, for the solution $\tilde{u}_h \in W_h$ of

$$a(\tilde{u}_h, \varphi - \tilde{u}_h) \geq (f, \varphi - \tilde{u}_h) \qquad \forall \varphi \in W_h \tag{9.4}$$

there holds $\tilde{u}_h = u_h$, i.e. u_h is equivalently characterised by (9.4). This inequality allows the choice of $I_h u$ on $\Omega \setminus B_h$ as a test function.

Consequently by localising the error to $\Omega \setminus B$ and $\Omega \setminus B_h$, the suboptimal term $\mathcal{O}(h^{1-2/p})$ only appears on a strip S consisting of cells $T \in \mathbb{T}_h$ along ∂B and ∂B_h respectively. In typical situations there clearly holds $\int_S dx = \mathcal{O}(h)$. This relation depends on the regularity of the solution, especially the structure of the *free boundary* (cf. Brézis and Stampacchia [18], Shaw [62]). Under this natural assumption, with $\int_S \mathcal{O}(h^{1-2/p}) dx = \mathcal{O}(h^{2-2/p})$, one one would obtain the estimates

$$\|\nabla(u - u_h)\|_{\Omega \setminus B}^2 = \mathcal{O}(h^{2-2/p}) \,, \qquad \|\nabla(u - u_h)\|_{\Omega \setminus B_h}^2 = \mathcal{O}(h^{2-2/p}) \,.$$

9.3   Saddle point problem

The construction of numerical algorithms for solving the discrete problems presented in the previous chapters is straightforward, since the restrictions only apply to the solutions itself. For example one can employ point projection schemes as described for the applications shown below. (cf. Glowinski et al. [34])

In contrast, in the present case, we have to deal with pointwise restrictions for the gradient of the solution. To handle this situation, we introduce the Lagrange functional

$$\mathcal{L}(\varphi, w) = \frac{1}{2} a(\varphi, \varphi) - (f, \varphi) + \int w((\nabla \varphi)^2 - 1) \tag{9.5}$$

for $\varphi \in V := H_0^1(\Omega)$ and $w \in \Lambda = \{q \in L^\infty \mid q \geq 0 \text{ a.e.}\}$.
If $\mathcal{L}(.,.)$ has a saddle point $(u, \lambda) \in V \times \Lambda$ – the existence is for example proven for constant f by Brézis [17] – it can be shown that the first component u

minimises the functional (9.1). In what follows, we assume f to be chosen in such a way, that $\mathcal{L}(.,.)$ has a solution in $V \times \Lambda$.

Choosing K_h and V_h as above, we introduce $\Lambda_h \subset \Lambda$ by

$$\Lambda_h = \{w \in \Lambda \mid w \text{ constant over each } T \in \mathbb{T}_h\}.$$

Proposition 3.5 in Glowinski [33] guarantees the existence of a saddle point $(u_h, \lambda_h) \in V_h \times \Lambda_h$ of the discrete analogue of (9.5). The first component u_h is then the solution of the original problem (9.3). Now, standard Uzawa-type schemes can be employed, to solve the discrete torsion problem (cf. Glowinski [33]).

Uzawa's scheme: The discrete problems of the are solved by the following iterative algorithm.

1. Choose an initial iterate λ_h^0 and $\rho > 0$

2. Solve the linear problem $\int (1 + \lambda_h^\nu) \nabla u_h \nabla \varphi = (f, \varphi) \; \forall \varphi \in V_h$

3. Update: $\lambda_h^{\nu+1} = \max(0, \lambda_h^\nu + \rho(|\nabla u_h^\nu|^2 - 1))$ on each cell.

4. Set $\nu = \nu + 1$ and go back to 2.

In addition, introducing this Lagrangian formalism allows for an improved error analysis of the torsion problem. For the calculation below, we set

$$b(\varphi, w) = \int w \nabla u \nabla \varphi,$$

$$s(\varphi, w) = \int w \nabla (u_h - u) \nabla \varphi,$$

$$b_h(\varphi, w) = b(\varphi, w) + s(\varphi, w),$$

$$g(w) = \int w,$$

$$A(v, \varphi) = a(v, \varphi) - (f, \varphi)$$

and state

Theorem 9.3.1. *With the notation introduced above, there holds the error estimate*

$$\|\nabla(u - u_h)\|^2 \le C(\|\nabla(\varphi_h - u)\|^2 + \|w_h - \lambda\|^2)$$
$$+ g(w_h - \lambda) - b(u, w_h - \lambda) - s(u_h, w_h) + \mathcal{O}(h^{2-2/p}) \qquad \forall \varphi_h \in V_h, \; w_h \in \Lambda_h.$$

Compared to the previous section, in practice, where $f \in L^\infty$, we have an improved convergence behaviour of the discretisation error measured in the energy norm tending to the order $\mathcal{O}(h)$. The proof is given in the following two subsections.

9.3.1   Discretisation error of the Lagrange multiplier

First, we show an error estimate for $\|\lambda - \lambda_h\|$. Since there holds

$$\mathcal{L}(u, \lambda) \leq \mathcal{L}(\varphi, \lambda) \quad \forall \varphi \in V\,, \qquad \mathcal{L}(u_h, \lambda_h) \leq \mathcal{L}(\varphi, \lambda_h) \quad \forall \varphi \in V_h\,,$$

we conclude

$$A(u, \varphi) + b(\varphi, \lambda) = 0 \quad \forall \varphi \in V\,, \tag{9.6}$$

$$A(u_h, \varphi) + b_h(\varphi, \lambda_h) = 0 \quad \forall \varphi \in V_h\,. \tag{9.7}$$

For arbitrary $\varphi \in V_h$ and $w \in \Lambda_h$ there holds

$$b(\varphi, w - \lambda_h) = b_h(\varphi, w) - b_h(\varphi, \lambda_h) - s(\varphi, w - \lambda_h)$$

use (9.6),(9.7)
$$= A(u_h, \varphi) - A(u, \varphi) - b(\varphi, \lambda) + b_h(\varphi, w) - s(\varphi, w - \lambda_h)$$

$$= A(u_h - u, \varphi) - b(\varphi, \lambda) + b(\varphi, w) + s(\varphi, w) - s(\varphi, w - \lambda_h)$$

$$= A(u_h - u, \varphi) + b(\varphi, w - \lambda) + s(\varphi, \lambda_h)$$

$$\leq \Big(\|\nabla(u - u_h)\| + c\|w - \lambda\| + c\|\lambda_h\|_{L^\infty} \|\nabla(u - u_h)\| \Big) \cdot \|\nabla\varphi\|\,.$$

Since the choice of our discrete spaces V_h and Λ_h fulfils the Babuška-Brezzi condition, we obtain

$$\|w - \lambda_h\| \leq C(\|\nabla(u - u_h)\| + \|w - \lambda\|) \tag{9.8}$$

and eventually we get the announced estimate for $\|\lambda - \lambda_h\|$

$$\|\lambda - \lambda_h\| \leq \|\lambda - w\| + \|w - \lambda_h\| \leq C(\|\nabla(u - u_h)\| + \|w - \lambda\|)\,, \tag{9.9}$$

for arbitrary $w \in \Lambda_h$.

Remark: We recall, that there holds the relation

$$((1 + \lambda_h)\nabla u_h, \nabla u_h) = \int_{\Omega \setminus B_h} (\nabla u_h)^2 \, dx + \int_{B_h} (1 + \lambda_h) \, dx = (f, u_h)\,,$$

from which we conclude $\|\lambda_h\|_{L^\infty}$ to be bounded.

9.3.2 Discretisation error of the primal variable

Next we derive an error estimate for $\|\nabla(u - u_h)\|$. Since there holds

$$\mathcal{L}(u, w) \leq \mathcal{L}(u, \lambda) \quad \forall w \in \Lambda, \qquad \mathcal{L}(u_h, w) \leq \mathcal{L}(u_h, \lambda_h) \quad \forall w \in \Lambda_h,$$

we conclude for $w \in \Lambda_h$

$$b(u, \lambda_h - \lambda) \leq g(\lambda_h - \lambda), \tag{9.10}$$

$$b(u_h, w - \lambda_h) \leq g(w - \lambda_h) - s(u_h, w - \lambda_h). \tag{9.11}$$

Now we estimate

$$
\begin{aligned}
b(u - u_h, \lambda_h - \lambda) &= b(u, \lambda_h - \lambda) - b(u_h, \lambda_h - \lambda) \\
\text{use (9.10)} \qquad &\leq g(\lambda_h - \lambda) - b(u_h, \lambda_h - w) - b(u_h, w - \lambda) \\
\text{use (9.11)} \qquad &\leq g(\lambda_h - \lambda) + g(w - \lambda_h) - s(u_h, w - \lambda_h) - b(u_h, w - \lambda)
\end{aligned}
$$

and summarise to

$$b(u - u_h, \lambda_h - \lambda) \leq g(w - \lambda) - b(u_h, w - \lambda) - s(u_h, w - \lambda_h). \tag{9.12}$$

Testing by $\varphi = u - u_h$ and $\varphi = u_h - u + u$ in (9.6) and (9.7), respectively, gives by adding these equations

$$
\begin{aligned}
0 = {} &A(u, u - u_h) + b(u - u_h, \lambda) + A(u_h, u_h - u + u) + b_h(u_h - u + u, \lambda_h) \\
&+ A(u - u_h, u - u_h) + A(u_h, u) + b(u - u_h, \lambda) + b(u_h - u + u, \lambda_h) + s(u_h, \lambda_h),
\end{aligned}
$$

eventually yielding

$$
\begin{aligned}
A(u - u_h, u - u_h) + b(u - u_h, \lambda - \lambda_h) & \\
+ A(u_h, u) + b(u, \lambda_h) + s(u_h, \lambda_h) &= 0. \tag{9.13}
\end{aligned}
$$

Using (9.13) and (9.12) one proceeds further with the estimate

$$
\begin{aligned}
A(u - u_h, u - u_h) &= b(u - u_h, \lambda_h - \lambda) - A(u_h, u) - b(u, \lambda_h) - s(u_h, \lambda_h) \\
&\leq g(w - \lambda) - b(u_h, w - \lambda) - s(u_h, w - \lambda_h) - A(u_h, u) - b(u, \lambda_h) - s(u_h, \lambda_h).
\end{aligned}
$$

Here the right-hand side can be estimated by employing (9.6) and (9.7) resulting in

$$
\begin{aligned}
g(w - \lambda) &- b(u_h, w - \lambda) - s(u_h, w - \lambda_h) - A(u_h, u) - b(u, \lambda_h) - s(u_h, \lambda_h) \\
= g(w - \lambda) &- b(u, w - \lambda) + b(u - u_h, w - \lambda) - A(u_h, u) - b(u, \lambda_h) - s(u_h, w) \\
&+ A(u, u - \varphi_h) + b(u - \varphi_h, \lambda) + A(u_h, \varphi_h) + b(\varphi_h, \lambda_h) + s(\varphi_h, \lambda_h) \\
= g(w - \lambda) &- b(u, w - \lambda) + b(u - u_h, w - \lambda) \\
&+ A(u - u_h, u - \varphi_h) + b(u - \varphi_h, \lambda - \lambda_h) - s(u_h, w) + s(\varphi_h, \lambda_h)
\end{aligned}
$$

Collecting the results of this subsection so far, the error estimate (9.9) for the dual variable and the stability argument,

$$
\|\nabla(u - u_h)\|^2 \le C A(u - u_h, u - u_h),
$$

gives the relation

$$
\begin{aligned}
\|\nabla(u - u_h)\|^2 \le C(&\|\nabla(\varphi_h - u)\|^2 + \|w - \lambda\|^2) + g(w - \lambda) \\
&- b(u, w - \lambda) - s(u_h, w) + s(\varphi_h, \lambda_h) \qquad \forall \varphi_h \in V_h, \ w \in \Lambda_h. \quad (9.14)
\end{aligned}
$$

This formally coincides with the result in Großmann and Roos [49] for bilinear $b(.,.)$ and $b_h(.,.) = b(.,.)$, up to the last two terms, which can be treated as follows.

$$
\begin{aligned}
- s(u_h, w) &+ s(\varphi_h, \lambda_h) \\
= &- \int w \nabla(u_h - u) \nabla u_h + \int \lambda_h \nabla(u_h - u) \nabla(\varphi_h - u + u) \\
\le &- \int w \nabla(u_h - u) \nabla u_h + \int \lambda_h \nabla(u_h - u) \nabla u \\
&+ \|\lambda_h\|_\infty \|\nabla(u_h - u)\| \ \|\nabla(\varphi_h - u)\|.
\end{aligned}
$$

Recalling the setting

$$
B = \{x \in \Omega \mid |\nabla u| = 1\}, \qquad B_h = \{T \in \mathbb{T}_h \mid |\nabla u_h| = 1\},
$$

we observe that on $B \cap B_h$ there holds

$$
- \int_{B \cap B_h} w \nabla(u_h - u) \nabla u_h \le 0,
$$

$$
\int_{B \cap B_h} \lambda_h \nabla(u_h - u) \nabla u \le 0.
$$

Using the remark from Section 2, the remaining part on $D = \Omega \setminus (B \cap B_h)$ is estimated by

$$- \int_D w \nabla(u_h - u) \nabla(u_h - u + u) + \int_D \lambda_h \nabla(u_h - u) \nabla u$$

$$= \underbrace{- \int_D w \nabla(u_h - u) \nabla(u_h - u)}_{\leq 0} + \int_D (\lambda_h - w) \nabla(u_h - u) \nabla u$$

$$\text{use (9.8)} \quad \leq \|\lambda_h - w\| \|\nabla(u_h - u)\|_D \|\nabla u\|_\infty$$

$$\leq \frac{1}{2} \|\nabla(u_h - u)\|^2 + \mathcal{O}(h^{2-2/p}),$$

which completes the proof of Theorem 9.3.1.

Chapter 10

Obstacle problem revisited

Motivated by the results of the previous chapter, we come back to the obstacle problem. Here, we treat our first example by the Lagrangian formalism and derive *a posteriori* error estimates for the corresponding FE-discretisation, employing the general framework from Chapter 8.

10.1 Weak formulation

With notation introduced in the chapter above, we define the Lagrange functional

$$\mathcal{L}(\varphi, w) = \frac{1}{2} a(\varphi, \varphi) - (f, \varphi) + \int_{\Omega} w(\psi - \varphi) \tag{10.1}$$

for $\varphi \in V := H_0^1(\Omega)$ and $w \in \Lambda = \{q \in L^2 \mid q \geq 0 \text{ a.e.}\}$, implying the weak formulation

$$a(u, \varphi) - (\lambda, \varphi) = (f, \varphi) \qquad \forall \varphi \in V,$$
$$(u, w - \lambda) \geq (\psi, w - \lambda) \qquad \forall w \in \Lambda,$$

for the obstacle problem, which can be written in compact form

$$a(u, \varphi) - (\lambda, \varphi) + (u, w - \lambda) \geq$$
$$(f, \varphi) + (\psi, w - \lambda) \quad \forall \{\varphi, w\} \in V \times \Lambda. \tag{10.2}$$

As a standard choice we take $V_h = \{\varphi \in H_0^1(\Omega) \mid \varphi \text{ linear on } T \in \mathbb{T}_h\}$ and L_h consisting of piecewise constant functions to state the discrete version of (10.2)

$$a(u_h, \varphi) - (\lambda_h, \varphi) + (u_h, w - \lambda_h) \geq (f, \varphi) + (\psi, w - \lambda_h), \tag{10.3}$$

for arbitrary $\{\varphi, w\} \in V_h \times \Lambda_h$ with $\Lambda_h = L_h \cap \Lambda$.

Uzawa's algorithm: The discrete problems of the form (10.3) are solved by an iterative algorithm of Uzawa-type.

1. Choose an initial iterate λ_h^0 and $\rho > 0$

2. Solve the linear problem $a(u_h^\nu, \varphi) = (f + \lambda_h^\nu, \varphi) \; \forall \varphi \in V_h$

3. Update: $\lambda_h^{\nu+1} = \max(0, \lambda_h^\nu + \rho(\psi - u_h^\nu))$ for each vertex of the FE-mesh.

4. Set $\nu = \nu + 1$ and go back to 2.

10.2   A posteriori error estimates

In order to apply the general concept, identifying U, U_h with the pairs $\{u, \lambda\}$, $\{u_h, \lambda_h\}$, the variational setting for the example under consideration is determined by

$$A(\{u, \lambda\}, \{\varphi, w\}) = a(u, \varphi) - (\lambda, \varphi) + (u, w),$$
$$F(\{\tau, \varphi\}) = (f, \varphi) + (\psi, w),$$
$$\mathbf{V} = \{\{\tau, \varphi\} \in V \times L^2\}, \qquad \mathbf{K} = \{\{\tau, \varphi\} \in V \times \Lambda\}.$$

In view of Lemma 8.1.2, we introduce the set $B_h = \{x \in \Omega \mid \lambda_h(x) = 0\}$. Now choosing $W_h = \{\{\varphi, w\} \in \mathbf{V}_h \mid w \geq 0 \text{ on } B_h\}$ the discrete problem (10.3) is equivalently defined by

$$a(u_h, \varphi) - (\lambda_h, \varphi) + (u_h, w - \lambda_h) \geq (f, \varphi) + (\psi, w - \lambda), \tag{10.4}$$

for arbitrary $\{\varphi, w\} \in W_h$. Together with $W_h^0 = W_h$ the assumptions of Lemma 8.1.2 are fulfilled.

Employing the proposed abstract framework for deriving *a posteriori* error bounds, we set

$$G = W^0 \cap \{\{\varphi, w\} \in \mathbf{V} \mid (\psi - u, w + \lambda_h - \lambda) \geq 0\},$$
$$W^0 = \{\{\varphi, w\} \in \mathbf{V} \mid w \geq 0 \text{ on } B_h\},$$

to formulate the dual problem with solution $(z, \pi) \in G$

$$J(\varphi, w - \pi) \leq a(\varphi, z) - (w - \pi, z) + (\varphi, \pi) \qquad \forall \{\varphi, w\} \in G. \tag{10.5}$$

Following the concepts of Chapter 8 this yields the *a posteriori* estimate

$$J(u - u_h, \lambda - \lambda_h) \leq$$
$$(f + \lambda_h, z - z_h) - a(u_h, z - z_h) + (\psi - u_h, \pi - \pi_h) \; \forall \{z_h, \pi_h\} \in W_h^0. \tag{10.6}$$

The approximation of G may be realised as follows. Recalling the complementarity condition $(u - \psi)\lambda = 0$, we assume $(u - \psi)\lambda_h \approx 0$. This gives the relation

$$(\psi - u, w + \lambda_h - \lambda) \approx (\psi - u, w) \geq 0. \tag{10.7}$$

Having in mind $\psi - u \leq 0$ together with requiring $w \geq 0$ on B_h this suggests to replace G by

$$G \approx \tilde{G} = \{\{\varphi, w\} \mid w = 0 \text{ on } B_h\}.$$

Remark: We recall the example of the *warning* in Chapter 5.

Consider an obstacle problem in one space dimension where ψ is chosen as the solution of the unrestricted problem. In this case, the corresponding FE-solution is simply the interpolant of ψ.

Our heuristic for the primal approach suggested the *a posteriori* error bound $|J(e)| \leq 0$, which is not true for arbitrary $J(\cdot)$.

In contrast, if we apply the strategy based on the Lagrangian formalism, we have $G \approx \tilde{G} = \{\{\varphi, w\} \mid w = 0 \text{ on } \Omega\}$. Based on the estimate (10.6), one gets

$$J(u - u_h, \lambda - \lambda_h) \leq (f + \lambda_h, z - z_h) - a(u_h, z - z_h) \qquad \forall \{z_h, \pi_h\} \in W_h^0.$$

We observe this estimate to be identical to the error estimate obtained by a duality argument for the unrestricted problem.

10.3   Numerical results

As a test example, we consider the obstacle problem on $\Omega = (0,1)^2$ with

$$f = 10.(x_1 - x_1^2 + x_2 - x_2^2), \qquad \psi(x) = -r^{3/2},$$

and $r = r(x) = (x_1 - 0.5)^2 + (x_2 - 0.5)^2$. As a functional, we choose $J(\varphi) = \varphi(x_0)$, $x_0 = (1/8, 1/4)$ to control the point-error in x_0.

The computational results are shown in Table 10.1. Evaluating *Ratio* shows the constant relation between the *true* error and the corresponding estimation. Consequently it is demonstrated that the proposed approach to *a posteriori* error control gives useful error bounds.

Cells	$J(u_h)$	E^{rel}	Ratio
64	-3.857922e-02	4.141504e-02	3.92
256	-3.770668e-02	1.786152e-02	1.84
1024	-3.725269e-02	5.606425e-03	1.52
4096	-3.711016e-02	1.758942e-03	1.26
16384	-3.705402e-02	2.434877e-04	1.52
65536	-3.704765e-02	7.153462e-05	1.47

Table 10.1: Numerical results for the test-example: functional value $J(u_h)$, relative error E^{rel} and over-estimation factor Ratio.

Remark: In our test calculations the proposed error estimator is evaluated as follows. The dual problem with the approximation $\tilde{G}$ is solved by the FE-method on a grid with local mesh-size $h_T/2$ obtained by a regular refinement step yielding the solution $\{z_{h/2}, \pi_{h/2}\}$.

Then the first and the last term in (10.6) are replaced by

$$(f + \lambda_h, z_{h/2} - r_h^1 z_{h/2}) \quad \text{and} \quad (\psi - u_h, \pi_{h/2} - r_h^0 \pi_{h/2})$$

respectively, where $(r_h^1, r_h^0) : \mathbf{V}_{h/2} \to \mathbf{V}_h$ denotes a standard restriction operator for the nested discrete spaces under consideration.

The second term is treated by

$$a(u_h, z - z_h) = (\nabla u_h, \nabla(z - z_h)) \approx (\nabla u_h, \mathcal{M}_{h/2} \nabla z_{h/2} - \nabla(r_h^1 z_{h/2})),$$

using the local averaging technique described above.

Chapter 11

Variational inequalities of second kind

In this chapter, we extend our studies on finite element Galerkin schemes for elliptic variational inequalities of first to the one of second kind. Especially we perform the corresponding *a posteriori* error analysis for a simple friction problem and a model flow of a Bingham fluid.

We focus on problems where $u \in V \subset H^1$ is sought fulfiling

$$a(u, \varphi - u) + j(\varphi) - j(u) \geq (f, \varphi - u) \qquad \forall \varphi \in V, \qquad (11.1)$$

$a(.,.)$ is assumed to be a positive definit bilinear form and the functional $j(\cdot)$ to be continuous and convex but not necessarily differentiable.

11.1 A flow problem

The physical problem under consideration is the special case of a steady laminar flow of a Bingham fluid in a cylindrical duct with cross-section Ω. Now, we seek for the flow velocity $u \in V := H_0^1$, which is characterised by (see e.g. Duvaut and Lions [26])

$$I(u) \leq I(\varphi) := \tfrac{1}{2} a(\varphi, \varphi) - (f, \varphi) + j(\varphi) \quad \forall \varphi \in V, \qquad (11.2)$$

with the definitions

$$a(u, \varphi) = \mu(\nabla u, \nabla \varphi), \quad j(\varphi) = g_0 \int_\Omega |\nabla \varphi| \, dx.$$

Here, f is the linear pressure drop and the constant $g_0 \geq 0$ describes a threshold below which the fluid behaves as a rigid material. Otherwise the behaviour is that of an incompressible fluid with viscosity $\mu > 0$.

Existence and uniqueness of the solution of the continuous and strictly convex functional $I(\cdot)$ follows from its coercivity (see e.g. Kinderlehrer and Stampacchia [44] or Glowinski et al. [34]).

Remark: u is also unique solution of the variational inequality (11.1). (cf. Glowinski et al. [34])

11.1.1   Saddle point problem

In Glowinski et al. [34] it is shown that the solution u of (11.2) is characterised by the existence of a Lagrange multiplier

$$\lambda \in \Lambda = \{ w \mid w \in L, \ |w(x)| \leq 1 \text{ a.e.} \}, \qquad L := (L^2(\Omega))^2,$$

such that the pair $\{u, \lambda\}$ is a saddle point on $V \times \Lambda$ of the Lagrangian

$$\mathcal{L}(\varphi, w) = \frac{1}{2} a(\varphi, \varphi) - (f, \varphi) + g(w, \varphi), \qquad g(w, \varphi) := g_0 \int_\Omega w \nabla \varphi \, dx.$$

This implies the weak formulation

$$a(u, \varphi) + g(\lambda, \varphi) = (f, \varphi) \qquad \forall \varphi \in V$$
$$-g(w - \lambda, u) \geq 0 \qquad \forall w \in \Lambda,$$

which can be written in compact form

$$a(u, \varphi) + g(\lambda, \varphi) - g(w - \lambda, u) \geq (f, \varphi) \quad \forall \{\varphi, w\} \in V \times \Lambda. \tag{11.3}$$

As a standard choice we take $V_h = \{\varphi \in H_0^1(\Omega) \mid \varphi \text{ linear on } T \in \mathbb{T}_h\}$ and L_h consisting of cellwise constant functions to state the discrete version of (11.3)

$$a(u_h, \varphi) + g(\lambda_h, \varphi) - g(w - \lambda_h, u_h) \geq (f, \varphi) \quad \forall \{\varphi, w\} \in V_h \times \Lambda_h, \tag{11.4}$$

where $\Lambda_h = L_h \cap \Lambda$.

Uzawa's algorithm: The discrete problems of the form (11.4) are solved by an iterative algorithm of Uzawa-type.

1. Choose an initial iterate λ_h^0 and $\rho > 0$

2. Solve the linear problem $a(u_h^\nu, \varphi) = (f, \varphi) - g(\lambda_h^\nu, \varphi) \; \forall \varphi \in V_h$

3. Update: $\lambda_h^{\nu+1} = P_\Lambda(\lambda_h^\nu + \rho \nabla u_h^\nu)$ on each cell.

4. Set $\nu = \nu + 1$ and go back to 2,

where the operator P_Λ is defined by $P_\Lambda(q) = q/\max(1, |q|)$, $q \in \mathbb{R}^2$.

11.2 A friction problem

A basic problem in the theory of elasticity with friction in classical notation reads

$$-\Delta u + u = f \qquad \text{in } \Omega, \tag{11.5}$$

with boundary conditions on Γ

$$|\frac{\partial u}{\partial n}| \leq g_0, \qquad u\frac{\partial u}{\partial n} + g_0|u| = 0, \tag{11.6}$$

with a material dependent constant $g_0 \geq 0$.

The basis for applying the FE-method to this problem is an inequality setting in the form (11.1)

$$a(u, \varphi - u) + j(\varphi) - j(u) \geq (f, \varphi - u) \qquad \forall \varphi \in V,$$

with the definitions

$$a(u, \varphi) = (\nabla u, \nabla \varphi) + (u, \varphi), \quad j(\varphi) = \int_\Gamma g_0|\gamma\varphi| \, d\Gamma.$$

where $\gamma : H^1(\Omega) \to L^2(\Gamma)$ denotes the standard trace operator.

Equivalently u is characterised by

$$I(u) \leq I(\varphi) := \tfrac{1}{2}a(\varphi, \varphi) - (f, \varphi) + j(\varphi) \quad \forall \varphi \in V. \tag{11.7}$$

Existence and uniqueness of the solution of the continuous and strictly convex functional $I(\cdot)$ follows from its coercivity (see e.g. Kinderlehrer and Stampacchia [44] or Glowinski et al. [34]).

11.2.1   Saddle point problem

In Glowinski et al. [34] it is shown that the solution u of (11.7) is characterised by the existence of a Lagrange multiplier

$$\lambda \in \Lambda = \{w \mid w \in L \,,\ |w(x)| \leq 1 \text{ a.e.}\}\,, \qquad L := L^2(\Gamma)\,,$$

such that the pair $\{u, \lambda\}$ is a saddle point on $V \times \Lambda$ of the Lagrangian

$$\mathcal{L}(\varphi, w) = \frac{1}{2}a(\varphi, \varphi) - (f, \varphi) + g(w, \varphi)\,, \qquad g(w, \varphi) := g_0 \int_\Gamma w(\gamma\varphi)\,d\Gamma\,.$$

This implies the weak formulation

$$a(u, \varphi) + g(\lambda, \varphi) = (f, \varphi) \qquad \forall \varphi \in V$$
$$-g(w - \lambda, u) \geq 0 \qquad \forall w \in \Lambda\,,$$

which can be written in compact form

$$a(u, \varphi) + g(\lambda, \varphi) - g(w - \lambda, u) \geq (f, \varphi) \quad \forall\{\varphi, w\} \in V \times \Lambda\,. \tag{11.8}$$

As a standard choice we take $V_h = \{\varphi \in H_0^1(\Omega) \mid \varphi \text{ linear on } T \in \mathbb{T}_h\}$ and L_h consisting of cellwise constant functions to state the discrete version of (11.8)

$$a(u_h, \varphi) + g(\lambda_h, \varphi) - g(w - \lambda_h, u_h) \geq (f, \varphi) \quad \forall\{\varphi, w\} \in V_h \times \Lambda_h\,, \tag{11.9}$$

where $\Lambda_h = L_h \cap \Lambda$.

Uzawa's algorithm: The discrete problems of the form (11.9) are solved by an iterative algorithm of Uzawa-type.

1. Choose an initial iterate λ_h^0 and $\rho > 0$

2. Solve the linear problem $a(u_h^\nu, \varphi) = (f, \varphi) - g(\lambda_h^\nu, \varphi)\ \forall \varphi \in V_h$

3. Update: $\lambda_h^{\nu+1} = P_\Lambda(\lambda_h^\nu + \rho\gamma u_h^\nu)$ on each line on the boundary.

4. Set $\nu = \nu + 1$ and go back to 2,

where the operator P_Λ is defined by $P_\Lambda(q) = q/\max(1, |q|)\,, q \in \mathbb{R}^2$.

11.3 A posteriori error estimate

In order to establish *a posteriori* estimates for the discretisation error measured in form of a (linear) functional $J(\cdot)$, we employ the duality argument given by Natterer [51].

We shorten the notation by defining

$$A(\{u,\lambda\},\{\varphi,w\}) = a(u,\varphi) + g(\lambda,\varphi) - g(w,u),$$
$$F(\{\varphi,w\}) = (f,\varphi),$$
$$\mathbf{V} = V \times (L^2(\Omega))^2, \quad \mathbf{K} = V \times \Lambda, \quad \mathbf{K}_h = V_h \times \Lambda_h.$$

Consequently by identifying U, U_h with the pairs $\{u,\lambda\}, \{u_h,\lambda_h\}$ the weak formulations for the example under consideration can be written in the form

$$A(U, \varphi - U) \geq F(\varphi - U) \qquad \forall \varphi \in \mathbf{K}, \tag{11.10}$$
$$A(U_h, \varphi - U_h) \geq F(\varphi - U_h) \qquad \forall \varphi \in \mathbf{K}_h. \tag{11.11}$$

Based on the experiences collected in Suttmeier [68], we consider the dual problem

$$J(\{\varphi, w - \pi\}) \leq a(\varphi, z) + g(w - \pi, z) - g(\pi, \varphi) \quad \forall \{\varphi, w\} \in G, \tag{11.12}$$

with dual solution $Z = \{z, \pi\} \in G$ and in a first run we take

$$G = \{ v \in \mathbf{V} \mid F(v + U_h - U) - A(U, v + U_h - U) \geq 0 \}$$
$$= \{ v \in \mathbf{V} \mid g(w + \lambda_h - \lambda, u) \geq 0 \}.$$

Choosing $\{u - u_h, \pi + \lambda - \lambda_h\}$ as a test function, we obtain

$$\begin{aligned}
J(U - U_h) &\leq A(U - U_h, Z) \\
&= A(U - U_h, Z - Z_h) + A(U - U_h, Z_h) \\
&\leq A(U - U_h, Z - Z_h) + A(U, Z_h + U_h - U) \\
&\quad - F(Z_h + U_h - U) + g(\pi_h, u_h) \\
&= A(U - U_h, Z - Z_h) + \underbrace{A(U, Z + U_h - U) - F(Z + U_h - U)}_{\leq 0} \\
&\quad + A(U, Z_h - Z) - F(Z_h - Z) + g(\pi_h, u_h), \\
&\leq F(Z - Z_h) - A(U_h, Z - Z_h) + g(\pi_h, u_h).
\end{aligned} \tag{11.13}$$

To get the second inequality (11.13), we use the estimate

$$A(U - U_h, Z_h) = a(u - u_h, z_h) + g(\lambda - \lambda_h, z_h) - g(\pi_h, u - u_h)$$

$$= \underbrace{(f, z_h) - a(u_h, z_h) - g(\lambda_h, z_h)}_{=0} + g(\pi_h, u_h)$$

$$- (f, z_h + u_h - u) + a(u, z_h + u_h - u) + g(\lambda, z_h + u_h - u) - g(\pi_h + \lambda_h - \lambda, u)$$

$$+ \underbrace{-g(\lambda - \lambda_h, u)}_{\leq 0} + \underbrace{a(u, u - u_h) + g(\lambda, u - u_h) - (f, u - u_h)}_{=0}$$

$$\leq A(U, Z_h + U_h - U) - F(Z_h + U_h - U) + g(\pi_h, u_h) \,.$$

Inserting our definitions, eventually we obtain

$$J(\{u - u_h, \lambda - \lambda_h\}) \leq (f, z - z_h) - a(u_h, z - z_h) - g(\lambda_h, z - z_h) + g(\pi, u_h) \,.$$

· This suggests an improved choice of G:

$$G = \{\{\varphi, w\} \in \mathbf{V} \mid g(w, u_h) \leq 0\}$$

$$\cap \{v \in \mathbf{V} \mid g(w + \lambda_h - \lambda, u) \geq 0\} \,. \quad (11.14)$$

Now, it remains to show $\{(u - u_h), \pi + \lambda - \lambda_h\}$ to be an admissible test function for the dual problem (11.12). To this end we proof

Lemma 11.3.1. $\{\varphi, \pi\} \in G$ *implies* $\theta = \{(u - u_h), \pi + \lambda - \lambda_h\} \in G$.

Proof.
i) With λ_i denoting the L^2-Projection, term II in

$$g(\pi + \lambda - \lambda_h, u_h) = g(\pi + \lambda - \lambda_i + \lambda_i - \lambda_h, u_h)$$

$$= \underbrace{g(\pi, u_h)}_{I} + \underbrace{g(\lambda - \lambda_i, u_h)}_{II} + \underbrace{g(\lambda_i - \lambda_h, u_h)}_{III} \leq 0$$

is zero. For $\{\varphi, \pi\} \in G$ term I is non-positive. Furthermore $(0, \lambda_i)$ is an admissible test function in the discrete variational formulation (11.9) which implies the term III to be negative or zero. Thus θ fulfils the first condition in G.

ii) Choosing $\varphi = U_h$ as a testfunction in the continuous variational inequality (11.10) and because $Z \in G$ there holds

$$0 \leq F(U - U_h) - A(U, U - U_h) + F(Z + U_h - U) - A(U, Z + U_h - U)$$

$$= F((Z + U - U_h) + U_h - U) - A(U, (Z + U - U_h) + U_h - U) \,.$$

Thus θ fulfils the second condition in G. $\square$

Summarising the above, we state

Theorem 11.3.1. *Let Z be the solution of the dual problem* (11.12) *on the set G defined in* (11.14), *then for the schemes* (11.4) *and* (11.9) *there holds the estimate*

$$J(\{u - u_h, \lambda - \lambda_h\})$$
$$\leq (f, z - z_h) - a(u_h, z - z_h) - g(\lambda_h, z - z_h) := \eta_{weight} \quad \forall z_h \in V_h. \quad (11.15)$$

In general the dual problem (11.12) cannot be solved analytically. In our test calculations we make use of the following approximation. Neglecting the difference $\lambda - \lambda_h$, we replace G by

$$\tilde{G} = \{\{\varphi, w\} \in \mathbf{V} \mid g(w, u_h) = 0\},$$

and solve the dual problem (11.12) on $\tilde{G}$ numerically on a grid with local mesh-size $h_T/2$ obtained by a regular refinement step yielding the solution $\{z_{h/2}, \pi_{h/2}\}$. Then instead of estimate (11.15) we evaluate

$$\eta_{weight} \approx (f, z_{h/2} - r_h^1 z_{h/2}) - a(u_h, z_{h/2} - r_h^1 z_{h/2}) - g(\lambda_h, z_{h/2} - r_h^1 z_{h/2}),$$

where $r_h^1 : V_{h/2} \to V_h$ denotes a standard restriction operator for the nested discrete spaces under consideration.

11.4 Numerical results

11.4.1 Bingham fluid

In order to check the approximation properties of the implementation of our scheme, we choose the following test problem on the circle $\Omega = \{x \in \mathbb{R}^2 \mid |x| \leq R := 1/2\}$, with $f = 0$, $\mu = 1$ and $g_0 = 1.25$. Then with $r = |x|$ and $R_S = (2g_0)/f$ the solution can be written in the form

$$u(r) = \begin{cases} \frac{R-r}{2}(\frac{c}{2}(R+r) - 2g_0) & r \geq R_S \\ \frac{R-R_S}{2}(\frac{c}{2}(R+R_S) - 2g_0) & \text{otherwise}. \end{cases}$$

Cells	Error	Red.
32	2.279000e-02	0.00
128	6.113554e-03	3.73
512	1.575329e-03	3.88
2048	3.958600e-04	3.98
8192	9.822265e-05	4.03

Table 11.1: Convergence behaviour measured in the L^2-norm with known solution.

The computations are done on a sequence of globally refined meshes evaluating the L^2-error of the displacements. The results are shown in Table 11.1. Evaluating the ratio *Red.* between the L^2-errors of two consecutive meshes indicates the expected convergence behaviour of $\mathcal{O}(h^2)$.

As a test for our error estimate, we choose $J(\varphi) = \varphi(x_0), x_0 = (1/4, 0)$ to control the point-error in x_0. The computational results are shown in Table 11.2. Evaluating *Ratio* shows the constant relation between the *true* error and the corresponding estimation. Consequently it is demonstrated that the proposed approach to *a posteriori* error control gives useful error bounds.

Cells	$J(u_h)$	E^{rel}	Ratio
32	1.637851e-01	4.822464e-02	3.83
128	1.570405e-01	5.059200e-03	1.33
512	1.558536e-01	2.536960e-03	2.43
2048	1.559304e-01	2.045440e-03	1.37
8192	1.561010e-01	9.536000e-04	1.62

Table 11.2: Numerical results for the test-example: functional value $J(u_h)$, relative error E^{rel} and over-estimation factor Ratio.

11.4.2 Friction problem

As a test configuration we consider (11.5) and (11.6) on $\Omega = (-1, 1)^2$ with $f = 2$ and $g_0 = 1$.

In order to investigate the behaviour of our error estimate, we choose $J(\varphi) = \varphi(x_0)$, $x_0 = (1/4, 0)$ to control the point-error in x_0. As a second test we take $J(\varphi) = \int_\Omega \varphi$ to estimate the error with respect to the mean value of u.

The computational results are shown in Tables 11.3 and 11.4. Evaluating *Ratio* shows the constant relation between the *true* error and the corresponding estimation. Consequently it is demonstrated that the proposed approach to *a posteriori* error control gives useful error bounds.

Cells	$J(u_h)$	E^{rel}	Ratio
128	2.571630e-01	1.485004e-02	2.22
512	2.539707e-01	2.252170e-03	2.76
2048	2.536270e-01	8.958169e-04	1.60
8192	2.534561e-01	2.213891e-04	1.48

Table 11.3: Numerical results for the test-example: functional value $J(u_h)$, relative error E^{rel} and over-estimation factor Ratio.

Cells	$J(u_h)$	E^{rel}	Ratio
32	5.674342e-01	2.014471e-02	1.57
128	5.763507e-01	4.747539e-03	1.70
512	5.780254e-01	1.855638e-03	1.04
2048	5.788618e-01	4.113279e-04	1.84
8192	5.791907e-01	1.566223e-04	1.24

Table 11.4: Numerical results for the second test-example: functional value $J(u_h)$, relative error E^{rel} and over-estimation factor Ratio.

Chapter 12

Time-dependent problems

In this chapter, we indicate how the abstract framework presented above for stationary problems may be appplied to non-stationary processes. This is illustrated at a parabolic model situation, which may be interpreted as a time-dependent obstacle problem,

$$
\begin{aligned}
\partial_t u - \Delta u - f &\geq 0\,, \\
u - \psi &\geq 0\,, \\
(u - \psi)(\partial_t u - \Delta u - f) &= 0\,,
\end{aligned}
\tag{12.1}
$$

defined on the space-time region $Q_T := \Omega \times I$, with time interval $I = (0, t_N)$. Furthermore the initial state is given by $u_{|t=0} = u_0$ and homogeneous boundary condition $u = 0$ are prescribed on $\partial\Omega$. Such problems arise in the field of *Computational Finance*, for instance for describing the behaviour of American options (see, e.g., Seydel [61]).

12.1 Discretisation

Setting $a(.,.) = (\nabla., \nabla.)$, the variational solution u with $u(0) = u_0$ is sought in $K = \{v \in V \mid u \geq \psi\}$ with

$$
V := H^1(I; L^2(\Omega)) \cap L^2(I; H_0^1(\Omega))
$$

satisfying

$$(\partial_t u, \varphi - u) + a(u, \varphi - u) \geq (f, \varphi - u), \quad \forall \varphi \in K. \qquad (12.2)$$

For discretisation, the time interval $[0, t_N]$ is decomposed like $0 = t_0 < t_1 < \ldots < t_N$ into subintervals $I_m := (t_{m-1}, t_m]$ of length $k_m := t_m - t_{m-1}$. Here, we only consider a variant of the simple Euler scheme which reads

$$(u^m - u^{m-1}, \varphi - u^m) + k_m a(u^m, \varphi - u^m)\, dt \geq (\bar{f}^m, \varphi - u^m)\, dt, \qquad (12.3)$$

with the averaged forcing terms $\bar{f}^m = \int_{I_m} f\, dt$.

At each time level t_m, we use finite element subspaces $V_h^m \subset V$, satisfying the conditions stated above. We emphasize that these spaces may vary in time in the course of mesh adaptation within each time step. Then, the discrete solution $u_h^m \in K_h^m := K \cap V_h^m$ is defined by

$$(u_h^m - u_h^{m-1}, \varphi - u_h^m) + k_m a(u_h^m, \varphi - u_h^m)\, dt \geq (\bar{f}^m, \varphi - u_h^m)\, dt, \qquad (12.4)$$

for all $\varphi \in K_h^m$.

We will perform the *a posteriori* analysis of the time-stepping error following the approach described in Erikson et al. [29]. To this end, we embed our solution scheme into a space-time Galerkin framework and use an associated space-time duality argument.

Corresponding to the sequences of values u_h^m at time t_m, we introduce the piecewise constant functions

$$u_h : [0, t_N] \to H_0^1(\Omega), \qquad u_h|_{I_m} \equiv u_h^m, \qquad u_h(0) := u_0.$$

Further, for functions w defined on I, we will employ the notation

$$w^m := w(t_m), \qquad w_\pm^m := \lim_{t \to 0} w(t_m \pm t), \qquad [w]^m := w_+^m - w_-^m,$$

and using settings introduced in Rannacher [53], we define

$$V_{h,k} := \{v \in L^\infty(I; H_0^1) \mid v(., t)_{|Q_T^n} \text{ is } d\text{-linear}, v(x, .)_{|Q_T^n} \text{ is constant}\},$$

where $Q_T^n := T \times I_n$. Notice, that $\partial_t u_h \equiv 0$ on I_m and $[u]^m = 0$. Then, the continuous solution u and the incremental discrete problems (12.4) can be written in variational form as

$$A_k(u, \varphi - u) \geq F_k(\varphi - u) \qquad \forall \varphi \in V,$$
$$A_k(u_h, \varphi - u_h) \geq F_k(\varphi - u_h) \qquad \forall \varphi \in V_{h,k},$$

with the bilinear form

$$A_k(v, \varphi) :=$$
$$\sum_{n=1}^{N} \int_{I_n} \{(\partial_t v, \varphi) + a(v, \varphi)\} dt + \sum_{n=2}^{N} ([v]^{n-1}, \varphi^{(n-1)+}) + (v^{0+}, \varphi^{0+}), \quad (12.5)$$

and the linear functional

$$F_k(\varphi) := \sum_{n=1}^{N} \int_{I_n} (f, \varphi) + (u_0, \varphi^{0+}). \quad (12.6)$$

12.2 Error estimation

Suppose, that the quantity to be controlled is the energy error $\|\nabla e(t_N)\|$ at the final time. For setting up a space-time duality argument, we define the sets

$$W_{h,k}^0 \subset \{z_h \in V_{h,k} \mid F_k(z_h) - A_k(u_h, z_h) \leq 0\},$$
$$G \subset \{v \in V \oplus V_{h,k} \mid F_k(v + u_h - u) - A_k(u, v + u_h - u) \geq 0\}$$

and consider the adjoint problem

$$A_k(\varphi - z, z) \geq$$
$$J(\varphi - z) := \|\nabla e^{N-}\|^{-1} (\nabla(\varphi - z)^{N-}, \nabla e^{N-}) \quad \forall \varphi \in G. \quad (12.7)$$

Then similar to the general result in Theorem 8.2.1, we obtain

Theorem 12.2.1. *The energy error* $J(e) = \|\nabla e(t_N)\|$ *at final time* t_N *is bounded by*

$$J(e) \leq F_k(z - z_h) - A_k(u_h, z - z_h) \quad \forall z_h \in W_{h,k}^0. \quad (12.8)$$

Chapter 13

Applications

The studies presented in this work are partly motivated by our membership in the research group *ForscherGruppe FreiFormFlächen* (FGFFF) [1] supported by the Deutsche Forschungsgemeinschaft. In collaboration with scientists from mechanical engineering, we provide the numerical analysis for problems arising in the field of highspeed machining. In what follows, we demonstrate the application of our strategies to grinding and milling processes. Furthermore we show some results concerning adaptive mesh design for FE models in elasto-plasticity.

13.1   Grinding

In this subproject, in a first run, a calibration of parameters of the physical model has to be performed. To this end we have to compare experiment (see Figure 13.1) and numerical simulation. The situation under consideration is as follows. A grinding disk is pressed onto a workpiece (see Figure 13.2). Now contact zone and associated normal forces have to be evaluated for experiment and simulation.

---

[1]http://www-isf.maschinenbau.uni-dortmund.de/fgfff/

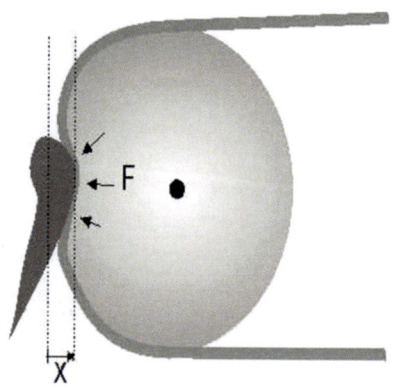

Figure 13.1: Snap shot of a grinding process.

Figure 13.2: Sketch of the model situation to be investigated for parameter calibration.

13.1.1   Discretisation

The underlying variational formulation for applying the FE-method to Signorini's problem (6.1) defines a weak solution $u \in K$ by

$$a(u, \varphi - u) \geq F(\varphi - u) \qquad \forall \varphi \in K, \tag{13.1}$$

with the definitions

$$V = \{v \in (H^1)^3 \,|\, v = 0 \text{ on } \Gamma_D\}, \qquad K = \{v \in V \,|\, v_n - g \leq 0 \text{ on } \Gamma_C\},$$

$$a(v, \varphi) = \int_\Omega A^{-1}\varepsilon(v)\varepsilon(\varphi) \quad \forall v, \varphi \in V, \qquad F(\varphi) = \int_\Omega f\varphi + \int_{\Gamma_N} t\varphi \quad \forall \varphi \in V.$$

The approximative solution $u_h \in K_h$ is determined by

$$a(u_h, \varphi - u_h) \geq F(\varphi - u_h) \qquad \forall \varphi \in K_h, \tag{13.2}$$

where we use standard trilinear finite elements to construct V_h.

In order to check the approximation properties of the implementation of our scheme, we choose the following test problem, restricted to linear elasticity: The constants in the material law $\sigma = 2\mu\varepsilon(u) + \lambda \operatorname{div} u$ are set to $\mu = 80000$

and $\lambda = 160000$. We employ zero boundary conditions for the displacements u_i on $\Omega = (0,1)^3$ and the prescribed right-hand side corresponds to

$$u_1 = (x^2 - 1)(y^2 - 1)(z^2 - 1)$$
$$\sin(2\pi x)\cos(4\pi yz),$$
$$u_2 = u_3 = 0.$$

The computations are done on a sequence of globally refined meshes evaluating the L^2-error of the displacements. The results are shown in Table 13.1. Evaluating the ratio $Red.$ between the L^2-errors of two consecutive meshes indicates the expected convergence behaviour of $\mathcal{O}(h^2)$.

Cells	DOF	Error	Red.
120	543	2.716+00	0.00
960	3531	7.016-01	3.87
7680	25491	1.794-01	3.90
61440	193827	4.508-02	3.98
491520	1512003	1.128-02	3.99

Table 13.1: Convergence behaviour measured in the L^2-norm with known solution.

13.1.2   Solution Process

The basic ingredient for the iterative solution process is a projected SOR-scheme introduced e.g. in Glowinski [33]. Let A denote the stiffness matrix associated to the discrete problem (13.2) and B, U the vectors of nodal values determining the discrete right-hand side and $u_h = u_h(U)$.

Now, under the constraints $u_h(U)(x) \cdot n \leq g$ for vertices $x \in \mathbb{T}_h$ with $x \in \Gamma_C$, we have to minimize $J(U) = U^T A U - 2 U^T B$.

Assuming $u_h(U^\nu) \cdot n \leq g$ on Γ_C, this is done by employing the iteration

1. Start by $U^{\nu+1/3} = A_{ii}^{-1} \left(B_i - \sum_{j<i} A_{ij} U_j^{\nu+1/3} - \sum_{j>i} A_{ij} U_j^{\nu} \right)$.

2. Update $U_i^{\nu+2/3} = U_i^{\nu} + \omega(U_i^{\nu+1/3} - U_i^{\nu})$ for $0 < \omega < 2$.

3. Project $U_i^{\nu+1} = P(U_i^{\nu+2/3})$, such that $u_h(U^{\nu+1}) \cdot n \leq g$ on Γ_C.

4. Set $\nu = \nu + 1$ and go back to 1.

Active set strategy

The convergence rate of the standard SOR-scheme deteriorates with decreasing mesh size. A remedy is to employ multilevel methods, which can be applied directly to the nonlinear problem as for example described in Hoppe and Kornhuber [37]. Alternatively and easier to implement if a linear multigrid solver is already available, are active set strategies (described e.g. in Erdmann et al. [27], [28]), i.e. an iteration between the nonlinear projection method and a linear multigrid scheme for adaptively refined meshes is constructed.

1. Choose an initial iterate u_h^0.

2. Perform a fixed number of steps (say 10) of the **nonlinear** SOR-scheme with initial iterate $u_h^{\nu-1}$ yielding $u_h^{\nu-1/2}$.

3. Set $B_h^\nu := \{x \in \Gamma_C \mid (u_h^{\nu-1/2} \cdot n)(x) = g(x)\}$.

4. Perform some steps of the **linear** multigrid-scheme with Dirichlet data on $\Gamma_D \cup B_h^\nu$
   with initial iterate $u_h^{\nu-1/2}$ yielding u_h^ν. The number of steps is chosen, such that the residual is decreased by a fixed ratio (say 0.1).

5. Set $\nu = \nu + 1$ and go back to 2.

13.1.3 Simulation

For the first test calculations, we assume isotropic linear elastic material behaviour, i.e., the constitutive relation between stresses and displacements is written

$$\sigma = 2\mu\varepsilon(u) + \lambda \operatorname{div} u \,,$$

with the material parameters $\mu = 1.02$ and $\lambda = 3.21$.

We apply an FE-strategy for the scheme (13.1) where the adaptive meshing is based on the *a posteriori* bound for the error measured in energy norm. In this case the error functional is given by $J(\cdot) = (A^{-1}\varepsilon(e), .)$. The corresponding dual solution is the error itself. Eventually we obtain the error bound

$$\|e\|_a^2 \leq \sum_{T \in \mathbb{T}_h} \omega_T \rho_T \tag{13.3}$$

with the local *residuals* ρ_T and *weights* ω_T defined by

$$\rho_T := h_T \| f + \mathrm{div}(A^{-1}\varepsilon(u_h)) \|_T + \tfrac{1}{2} h_T^{1/2} \| n \cdot [A^{-1}\varepsilon(u_h)] \|_{\partial T},$$
$$\omega_T := \max \left\{ h_T^{-1} \| e - e_h \|_T, h_T^{-1/2} \| e - e_h \|_{\partial T} \right\},$$

where $[A^{-1}\varepsilon(u_h)]$ denotes the jump of $A^{-1}\varepsilon(u_h)$ across the interelement boundary.

Now in order to evaluate the weights ω_T, one can use the interpolation estimate (see, e.g., Hansbo and Johnson [41])

$$\omega_T \le C_{i,T} \| e \|_{a,\tilde{T}},$$

where $\tilde{T}$ is the union of T and its neighbours, establishing

$$\| e \|_a \le c \Big(\sum_{T \in \mathbb{T}_h} C_{i,T}^2 \rho_T^2 \Big)^{1/2}.$$

Alternatively, using the notation $\sigma(\cdot) = A^{-1}\varepsilon(\cdot)$, the result (13.3) can be estimated by

$$\| e \|_a^2 \le \sum_{T \in \mathbb{T}_h} C_{i,T} \rho_T \left(\int_{\tilde{T}} (\sigma(u) - \sigma(u_h))(\varepsilon(u) - \varepsilon(u_h)) \right)^{1/2}.$$

The further procedure is based on the idea of higher–order recovery of the derivatives of u by local averaging according to the concept of Zienkiewicz and Zhu [74] . The local weight $\| u - u_h \|_{a,\tilde{T}}$ is thought to be well represented by the auxiliary quantity

$$\tilde{\omega}_T^2 := \int_{\tilde{T}} (\mathcal{M}_h \sigma(u_h) - \sigma(u_h))(\mathcal{M}_h \varepsilon(u_h) - \varepsilon(u_h)),$$

where $\mathcal{M}_h \nabla u_h$ is a local (super-convergent) approximation of ∇u. Eventually, we get the weighted approximate *a posteriori* error bound

$$\| e \|_a^2 \approx \sum_{T \in \mathbb{T}_h} C_{i,T} \tilde{\omega} \rho_T.$$

In Figure 13.3 there is depicted a sequence of locally refined grids produced by our numerical simulation, showing that especially the contact zone is well resolved.

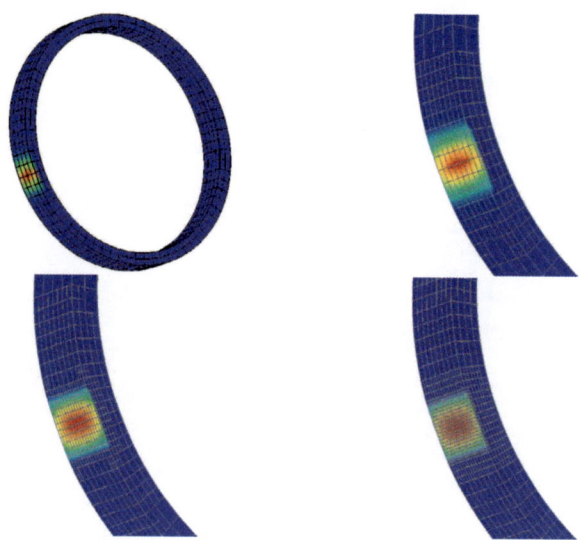

Figure 13.3: Sequence of locally refined grids (with zoom to contact zone) produced by our numerical simulation, showing that especially the contact zone is well resolved.

In addition we investigate the behaviour of the iterative solution process introduced above. The calculations are done on a SUN workstation. We compare the solution times required for solving the discrete systems on adaptively refined grids for the projected relaxation method SOR and the active set iteration SOR+, where the time for SOR+ on the coarsest mesh is set to one. Figure 13.4 and Table 13.2 show SOR to suffer from rapidly increasing solution times, whereas SOR+ turns out to give a significant speed up *Rate* for solving.

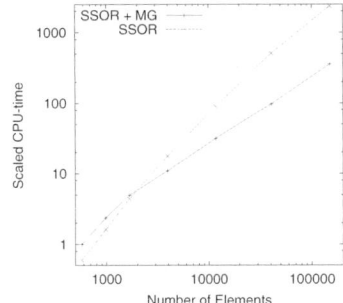

	Scaled CPU-time		
Cells	SOR+	SOR	Rate
576	1.0	0.5	0.59
982	2.3	1.6	0.68
1668	4.9	4.3	0.88
3957	10.9	17.7	1.61
11713	31.5	90.0	2.85
40840	97.6	508.1	5.20
149284	356.8	2362.3	6.61

Figure 13.4: Computing times for SOR and SOR+, demonstrating SOR to suffer from rapidly increasing solution times.

Table 13.2: Speed up between SOR and SOR+, showing the last to give a significant improvement.

13.2 Milling

In this section the underlying situation is depicted in Figures 13.5 and 13.6. As done in the previous subsection, we show in Figure 13.7 a sequence of locally refined grids produced by our numerical simulation, showing that especially the contact zone is well resolved.

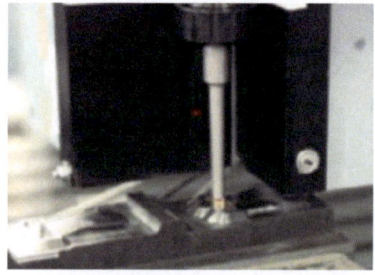

Figure 13.5: Snap shot of a milling process.

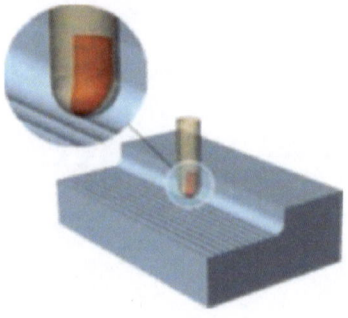

Figure 13.6: Sketch of model situation of a milling process.

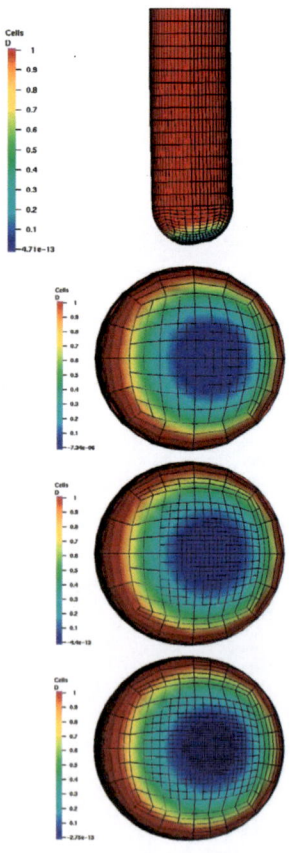

Figure 13.7: Sequence of locally refined grids (with zoom to contact zone) produced by our numerical simulation, showing that especially the contact zone is well resolved.

13.3 Elasto-plastic benchmark problem

In order to demonstrate our techniques to work within a time-stepping scheme on locally varying meshes in space including hanging nodes, we show some results concerning this subject partly taken from Suttmeier [66] and Rannacher and Suttmeier [58].

The fundamental problem in the flow theory of linear–elastic perfectly plastic material in classical notation reads (see, e.g., [26])

$$
\begin{aligned}
-\operatorname{div}\sigma = f, \quad \varepsilon(\dot{u}) &= A\dot{\sigma} + \lambda \qquad \text{in } \Omega, \\
\lambda(\tau - \sigma) \leq 0 \quad \forall\,\tau \text{ with } \mathcal{F}(\tau) \leq 0, \quad \lambda\dot{\sigma} &= 0 \qquad \text{in } \Omega, \\
\dot{u} = 0 \ \text{on } \Gamma_D, \qquad \sigma \cdot n &= g \ \text{on } \Gamma_N,
\end{aligned}
\tag{13.4}
$$

where σ and u are the stress tensor and displacement vector, respectively. We assume a stress-free initial state $\sigma(0) = 0$ and $u(0) = 0$. This idealized model describes the deformation of an elasto-plastic body occupying a bounded domain $\Omega \subset \mathbb{R}^d$ ($d = 2$ or 3) under the action of a body force f and a surface traction g along Γ_N. Along the remaining part of the boundary, $\Gamma_D = \partial\Omega \setminus \Gamma_N$, the body is fixed. The plastic growth is denoted by λ, and $\mathcal{F}(\cdot)$ is the (convex) von Mises yield function. We assume a quasi-static process, i.e., acceleration effects are neglected. Accordingly, the loading is prescribed in terms of functions $f(x,t) = a(t)f_0(x)$ and $g(x,t) = b(t)g_0(x)$ depending on a time-like parameter $t \in I := [0, T]$, where the functions $a(t)$ and $b(t)$ vary only slowly (relative to the elasto-plastic changes in the model). Further, the deformation is supposed to be small, so that the strain tensor can be written as $\varepsilon(u) = \frac{1}{2}(\nabla u + \nabla u^T)$. The material tensor A is assumed to be symmetric and positive definite.

The problem (13.4) is to be solved by the finite element Galerkin method on adaptively optimized meshes. For applying a finite element method, we have to rewrite problem (13.4) in a variational setting. To this end, we introduce the function spaces

$$
H := L^2(\Omega, \mathbb{R}^d), \qquad W := L^2(\Omega, \mathbb{R}^{d \times d}_{sym}), \qquad W^{div} := \{\tau \in W,\ \operatorname{div}\tau \in H\},
$$
$$
W^{div}_g := \{\tau \in W^{div},\ \tau \cdot n = g \text{ on } \Gamma_N\},\ W^{div}_{f,g} := \{\tau \in W^{div}_g,\ -\operatorname{div}\tau = f \text{ in } \Omega\},
$$
$$
V := \left\{v \in H^1(\Omega, \mathbb{R}^d),\ v = 0 \text{ on } \Gamma_D\right\},
$$

and define for any such space Σ the *admissible set* as

$$\Pi\Sigma := \{\tau \in \Sigma, \mathcal{F}(\tau) \leq 0\}.$$

Further, $(\cdot, \cdot)$ and $\|\cdot\|$ denote the L^2-inner product and norm over Ω, and $(\cdot, \cdot)_{\Gamma_N}$ is the L^2-inner product over the curve segment Γ_N.

Following Duvaut & Lions [26], we introduce the displacement velocity $v := \dot{u}$ and state the *dual-mixed* formulation of (13.4): Find a pair $\{v, \sigma\} : I \rightarrow H \times \Pi W_g^{div}$, with $\sigma(0) = 0$, $v(0) = 0$, satisfying

$$(A\dot{\sigma}, \tau - \sigma) + (v, \operatorname{div}\tau - \operatorname{div}\sigma) - (\operatorname{div}\sigma, \varphi) \geq (f, \varphi) \tag{13.5}$$

for arbitrary $\{\varphi, \tau\} \in H \times \Pi W_0^{div}$. Integrating by parts in (13.5) leads to the *primal-mixed* variational formulation: Find a pair $\{v, \sigma\} : I \rightarrow V \times \Pi W$, with $\sigma(0) = 0$, $v(0) = 0$, satisfying

$$(A\dot{\sigma} - \varepsilon(v), \tau - \sigma) + (\sigma, \varepsilon(\varphi)) \geq F(\varphi) \qquad \forall \{\varphi, \tau\} \in V \times \Pi W, \tag{13.6}$$

where the righthand side has the form $F(\varphi) := (f, \varphi) + (g, \varphi)_{\Gamma_N}$.

With notations introduced above, at each time level t_m, we use the subspaces $V_h^m \times W_h^m \subset V \times W$, where we consider standard finite element subspaces $V_h^m \subset V$ and $W_h^m \subset W$, and confine ourselves to the lowest-order approximation by continuous P_1- or Q_1-elements (linear or (iso-parametric) d-linear shape functions) for the deformations in V_h, while the corresponding discrete stresses in W_h are chosen as discontinuous P_0- or likewise Q_1-elements, respectively.

Then the discrete velocities and stresses $\{v_h^m, \sigma_h^m\} \in V_h^m \times \Pi W_h^m$ are defined by

$$(A\sigma_h^m - A\sigma_h^{m-1} + k_m \varepsilon(v_h^m)), \tau - \sigma_h^m) + k_m(\sigma_h^m, \varepsilon(\varphi)) \geq k_m F^m(\varphi), \tag{13.7}$$

for all $\{\varphi, \tau\} \in V_h^m \times \Pi W_h^m$, starting from the initial stress $\sigma_h^0 = 0$.

We apply our adaptive approach to a standard benchmark problem. A geometrically two-dimensional square disc with a hole is subjected to a constant boundary traction acting upon two opposite sides. No body force is applied, i.e., $f \equiv 0$. In the 2-dimensional case we use the plane-strain approximation and assume perfectly plastic material behaviour. In virtue of symmetry the consideration can be restricted to a quarter of the domain as shown in Figure 13.8. The height and width of the quarter corresponding to lines $\overline{45}$ and $\overline{15}$ are 100, and the radius of the hole is 10. The ratio of the lines $\overline{57}$ and $\overline{67}$ is $2:1$. The boundary traction is given in the form

$$g(t) = t g_0, \quad g_0 = 100, \quad t \in [0, 4.5].$$
(13.8)

We consider the case of a linear–elastic isotropic material law expressed in the form

$$\sigma = C\varepsilon := 2\mu\varepsilon^D(u) + \kappa \operatorname{div} u \, I,$$

with material dependent constants $\mu > 0$ and $\kappa > 0$. Here, $C = A^{-1}$, and $\tau^D := \tau - \frac{1}{3}tr(\tau)I$ is the deviatoric part of a tensor τ. The von Mises flow function has the form

$$\mathcal{F}(\sigma) = |\sigma^D|^2 - \sigma_0^2 \leq 0.$$

In our test calculations, we have chosen the material parameters as commonly used for aluminium: $\kappa = 164,206 \, N/mm^2$, $\mu = 80,193.80 \, N/mm^2$, $\sigma_0 = \sqrt{2/3}\,450$.

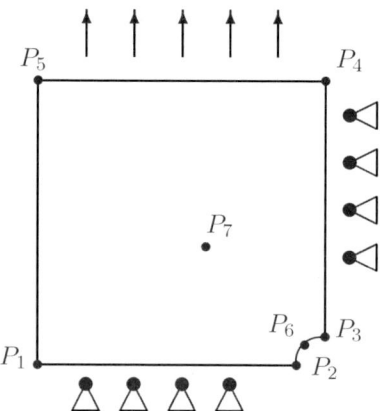

Figure 13.8: Geometry of the benchmark problem.

13.3.1   Solution of the benchmark: Hencky model

Assuming Hencky material behaviour, the calculations are performed with one load step from $t = 0$ to $t = 4.5$. The quantities to be computed are:

- stress σ_{22} and displacement u_1 at point P_2,

- Displacement u_1 at point P_5,

- Displacement u_2 at point P_4,

- Line integral $L_\Gamma(u_2)$ of displacement u_2 along $\Gamma = \overline{45}$,

- L^2-norm $\|\sigma\|_\Omega$ over Ω.

We present the computational results in the following way: In each subsection, the table demonstrates the sharpness of the *weighted* a posteriori error bound $\eta_\omega(u_h)$, while the figure indicates the superior efficiency of the generated meshes compared to those obtained by using the *energy* and the ZZ-error indicators, $\eta_E(u_h)$ and $\eta_{ZZ}(u_h)$, respectively. Finally, the adaptive grids generated by weighted error estimator with about $10,000$ cells are displayed.

Computation of $\sigma_{22}(P_2)$ (Hencky model)

N	$\sigma_{22}(P_2)$	E_ω^{rel}	Ratio
1000	5.9654e+02	1.4804e-01	2.0553e+02
2000	5.5653e+02	7.1025e-02	1.2828e+02
4000	5.3617e+02	3.1857e-02	6.6601e+01
8000	5.2330e+02	7.0786e-03	1.8520e+01
16000	5.1978e+02	3.0426e-04	2.3218e+00
∞	5.1962e+02		

Table 13.3: Results for $\sigma_{22}(P_2)$ based on the error estimator η_ω

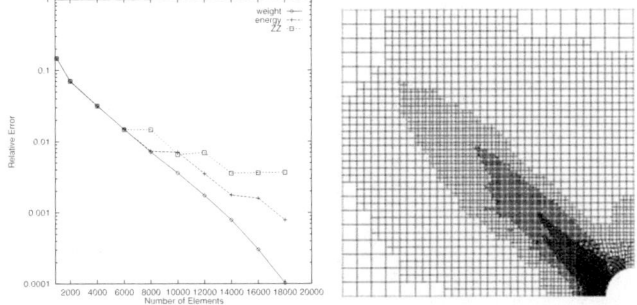

Figure 13.9: Relative error for computation of $\sigma_{22}(P_2)$ based on the different error indicators and an "optimal" mesh with about 10,000 cells

Computation of $u_1(P_2)$ (Hencky model)

N	$u_1(P_2)$	E_ω^{rel}	Ratio
1000	9.3081e-03	4.5567e-01	3.0157e+00
2000	9.5409e-03	4.4205e-01	2.1351e+00
4000	1.0552e-02	3.8291e-01	1.4675e+00
8000	1.3524e-02	2.0912e-01	1.2240e+00
16000	1.4925e-02	1.2718e-01	1.4689e+00
∞	1.7104e-02		

Table 13.4: Results for $u_1(P_2)$ based on the error estimator η_ω

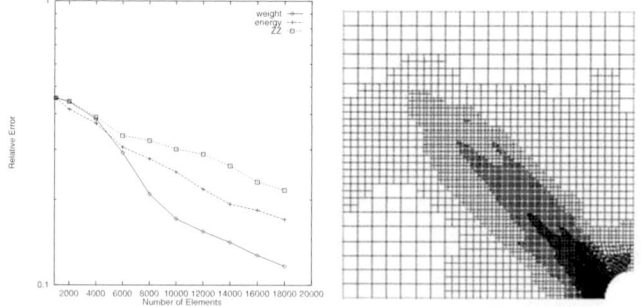

Figure 13.10: Relative errors for computation of $u_1(P_2)$ based on the different estimators and an "optimal" mesh with about 10,000 cells

Computation of $u_1(P_5)$ (Hencky model)

N	$u_1(P_5)$	E_ω^{rel}	Ratio
1000	6.5991e-02	7.7403e-02	6.4739e-01
2000	6.3462e-02	3.6121e-02	8.3647e-01
4000	6.2159e-02	1.4846e-02	1.0499e+00
8000	6.1554e-02	4.9704e-03	1.5502e+00
16000	6.1389e-02	2.2746e-03	1.7415e+00
∞	6.1251e-02		

Table 13.5: Results for $u_1(P_5)$ based on the error estimator η_ω

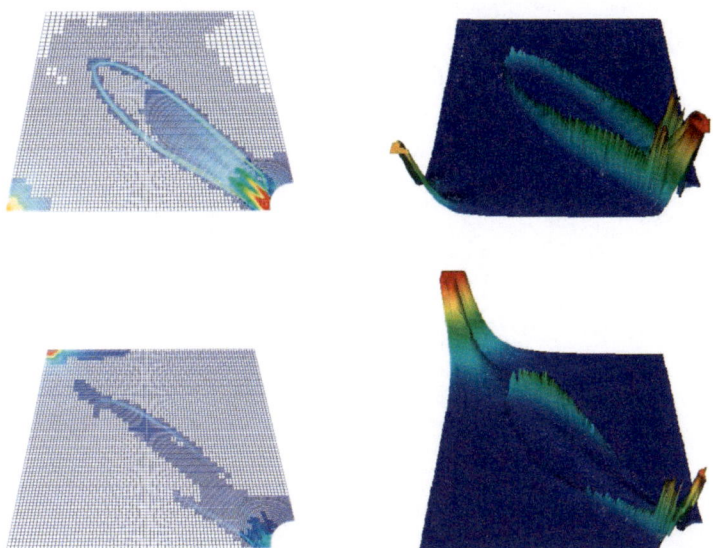

Figure 13.11: Optimized meshes for computing $u_1(P_1)$ (top) and $u_1(P_5)$ (bottom) together with corresponding weight distributions ω_T.

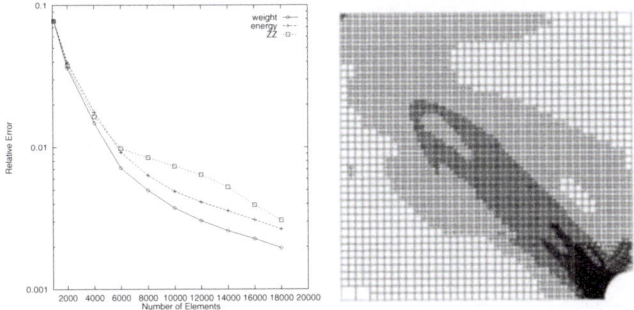

Figure 13.12: Relative error for computation of $u_1(P_5)$ based on the different error indicators and an "optimal" mesh with about 10,000 cells

Figure 13.11 shows optimized meshes with $N \approx 10000$ cells and the distributions of the corresponding weights ω_T for computing $u_1(P_1)$ and $u_1(P_5)$. We clearly see a strong effect of the structure of the dual solution on the resulting mesh refinement in the plastic zone. For representing the value $u_1(P_1)$ with good accuracy a good resolution of the whole elastic-plastic transition zone is required, while the representation of $u_1(P_5)$ poses less requirements in this respect. The resulting significant differences in the mesh-size distributions could have hardly be predicted by a priori considerations. Clearly, in this test case the meshes generated by the DWR-method are significantly more efficient for computing different target quantities than those obtained by the ZZ or energy-norm error indicators. This effect is especially pronounced when high accuracy is required.

Computation of $u_2(P_4)$ (Hencky model)

N	u_2^4	E_ω^{rel}	Ratio
1000	2.3568e-01	4.6994e-02	6.5408e-01
2000	2.4192e-01	2.1766e-02	8.2389e-01
4000	2.4509e-01	8.9232e-03	1.0034e+00
8000	2.4652e-01	3.1468e-03	1.3763e+00
16000	2.4690e-01	1.6122e-03	1.3480e+00
∞	2.4730e-01		

Table 13.6: Results for $u_2(P_4)$ based on the error estimator η_ω

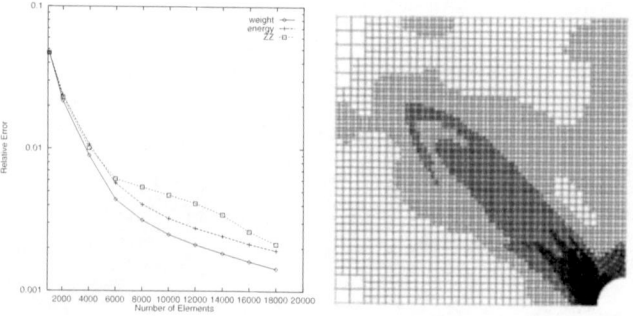

Figure 13.13: Relative error for computation of $u_2(P_4)$ based on the different error indicators and an "optimal" mesh with about 10,000 cells

Computation of line integral $L_\Gamma(u_2)$ over $\Gamma = \overline{45}$ (Hencky model)

N	$L_\Gamma(u_2)$	E_ω^{rel}	Ratio
1000	2.1760e+01	3.0944e-02	6.0564e-01
2000	2.2126e+01	1.4642e-02	7.7506e-01
4000	2.2318e+01	6.0949e-03	9.5711e-01
8000	2.2410e+01	2.0214e-03	1.4112e+00
16000	2.2433e+01	9.9800e-04	1.4671e+00
∞	2.2454e+01		

Table 13.7: Results for $L_\Gamma(u_2)$ based on the error estimator η_ω

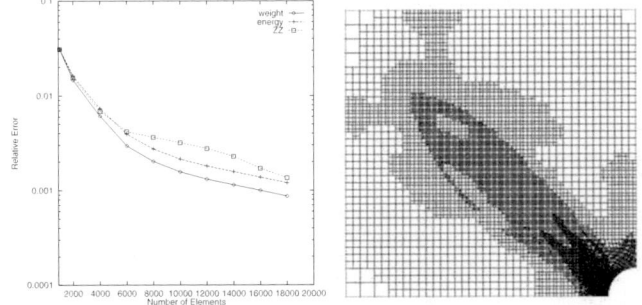

Figure 13.14: Relative error for computation of $L_\Gamma(u_2)$ based on the different error indicators and an "optimal" mesh with about 10,000 cells

Computation of norm $\|\sigma\|_\Omega$ over Ω (Hencky model)

N	$\|\sigma_h\|_\Omega$	E_ω^{rel}	Ratio
1000	1.702970e+10	5.4505e-03	2.9646e+00
2000	1.708084e+10	2.4638e-03	2.6479e+00
4000	1.711008e+10	7.5642e-04	2.1612e+00
8000	1.711481e+10	4.7986e-04	1.5581e+00
16000	1.711963e+10	1.9868e-04	1.1283e+00
∞	1.712303e+10		

Table 13.8: Results for $\|\sigma\|_\Omega$ based on the error estimator η_ω

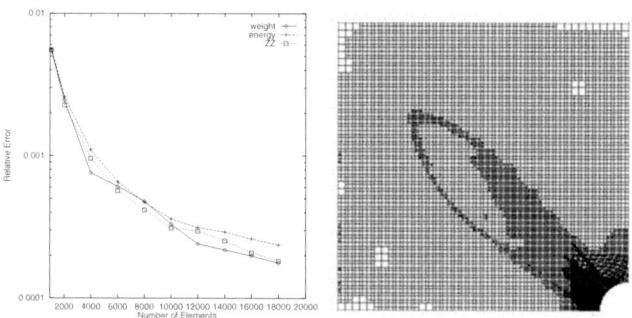

Figure 13.15: Relative error for computation of $\|\sigma\|_\Omega$ based on the different error indicators and an "optimal" mesh with about 10,000 cells

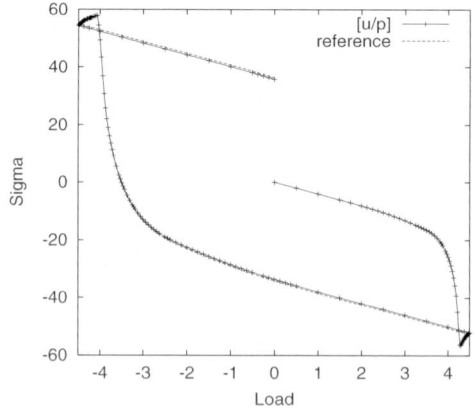

Figure 13.16: Plot of the reference solution (500.000 DOF) and the one obtained by our adaptive scheme (12.000 DOF).

13.3.2   Solution of the benchmark: Prandtl-Reuss model

We evaluate values of the stresses in point P_7 under cycling loading. We compare our discretisation, which uses an *a posteriori* error estimate controlling the errors in time and space (cf. [57]) to a *reference solution*, produced on the basis of a primal formulation approximating the displacements by the continuous, standard (bi)-linear shape functions. The pressure is treated by an inner variable (see [73]). About 500.000 degrees of freedom (DOF) were employed to guarentee an error below one percent. In Figure 13.16, there is shown the plot of the reference solution and the one obtained by our adaptive scheme using only about 12.000 DOF to reach the prescribed accuracy.

Further, Figure 13.17 shows a sequence of zooms into adaptively refined meshes for computing $\sigma_1 1(P_7)$ over the loading path $3.6 \leq t \leq 4.2$.

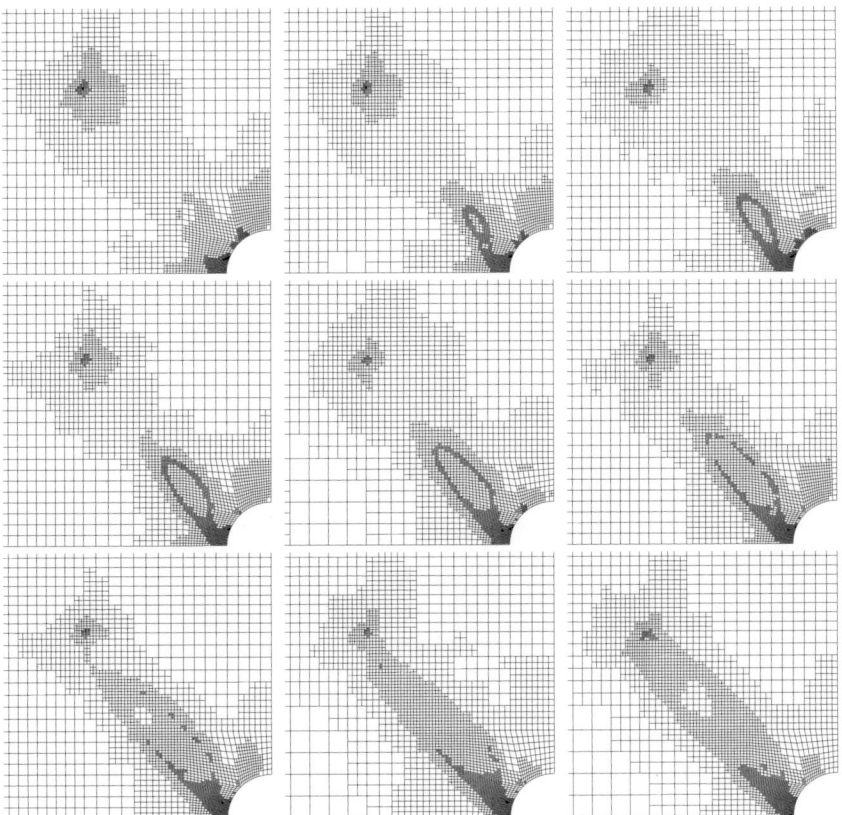

Figure 13.17: Sequence of the locally refined meshes including hanging nodes. Benchmark problem with zoom to one quarter of the configuration.

Chapter 14

Iterative Algorithms

In this chapter, we present ideas and recent results for a cascadic multigrid algorithm for variational inequalities.

When classical multigrid methods are applied to discretisations of variational inequalities frequently several complications are encountered mainly due to the lack of simple feasible restriction operators. These difficulties vanish in the application of the cascadic version of the multigrid method which in this sense yields greater advantages than in the linear case. Furthermore, a cg-method is proposed as smoother and as solver on coarse meshes. The efficiency of the new algorithm is elucidated by test calcuations for an obstacle problem and for a Signorini problem.

14.1 Introduction

As indicated throughout this book, a typical example for studying variational inequalities is

$$\int_\Omega [\frac{1}{2}(\nabla v)^2 - fv]dx \rightarrow \min! \tag{14.1}$$

with the boundary conditions $v(x) = g(x)$ on $\Gamma_D \subset \partial\Omega$ and the one-sided restrictions

$$v(x) \geq \psi(x) \quad \text{for} \quad x \in \Omega_\geq.$$

Here $\Omega_{\geq}$ is either a subset of Ω or of $\partial\Omega$. Formally we are looking for a solution in a closed convex cone in $H^1(\Omega)$ and not in a linear space.

There is a natural interest to solve the inequalities which arise from a finite element discretisation by the multigrid method. Unfortunately, there is a complication that is not encountered in linear problems. Let v_h^0 be the actual approximation in our multigrid iteration at the level with mesh size h. Then we are looking for the solution of a minimization problem in a set $v_h^0 + K_h$, where K_h is a convex cone whose boundary is given by a finite element function on the mesh with mesh size h. When we proceed to a coarser grid, we have also to coarsen the finite element function that specifies the cone. As a consequence, we get an approximation of the cone (in addition to the reduction of the dimension).

In the eighties (of the last century) there were used either inner approximations of the cone, cf. Hackbusch and Mittelmann [35], or outer approximations, cf. Brandt and Cryer [16] or Bollrath [14]. The inner approximation can lead to so small cones that the multigrid efficiency gets lost, and the (outer) approximation by a larger cone requires that a projection onto the feasible set is added, and multigrid efficiency can also not be guaranteed. These disadvantages were avoided by Kornhuber [45] who modified the basis functions at the frontier of the active set, cf. also [37].

We will take a different approach and use a *cascadic multigrid algorithm* avoiding the return to coarser grids in the iteration. Cascadic algorithms have been used successfully for linear problems [15, 22], and the advantage to employ transfer operators between the grids in one direction only, is here even clearer than in the application to linear elliptic problems.

The crucial point of a successful usage of cascadic codes is that the number of smoothing steps is sufficiently large and well balanced on the coarser grids [22]. This feature is also essential when treating variational inequalities and makes the difference to the code in [12] where the computations on coarse grids were restricted to the subdomain of inactive points.

The cascadic iteration starts on the coarsest mesh, and there an iteration of cg-type is applied. We note that a different cg code was proposed by Dostál and Schöberl [25]. The iteration step in their cg-code as well as in our one is more expensive than in the linear case since a possible change of the set of active points implies that the update of the gradients requires an extra matrix-vector-multiplication. Nevertheless, we recommend cg-steps as solver on the coarsest level as well as in the smoothing procedures at the higher levels. The efficiency of this approach and the whole cascadic multigrid iteration will be documented by several twodimensional numerical examples.

14.2 A Smoothing Procedure

The solution u of the variational problem (14.1) is characterized by the inequalities

$$(\nabla u, \nabla(v - u)) \geq (f, v - u) \qquad \forall v \in K, \tag{14.2}$$

where

$$K := \{v \in H^1(\Omega); \ v \geq \psi \text{ for } x \in \Omega_\geq \text{ and } v = g \text{ on } \Gamma_D\},$$

and $(.,.)$ denotes the inner product in $L_2(\Omega)$.

With notations introduced above, the finite element solution u_h belongs to the induced cone

$$K_h := \{v \in V_h; \ v(x) \geq \psi_h \text{ for } x \in \Omega_\geq\}$$

and is characterized by

$$(\nabla u_h, \nabla(v - u_h)) \geq (f, v - u_h) \qquad \forall v \in K_h. \tag{14.3}$$

Since (14.3) describes the minimum of an elliptic expression (function) on a convex set, the solution of (14.3) is unique.

When the solution is determined, a nodal basis is used. For convenience, we identify a finite element function $v \in V_h$ with the vector v whose components are the nodal values

$$v_i := v(x_i).$$

Similarly, we have $\psi_i := \psi(x_i)$, defining the interpolant $\psi_I \in V_h$ of ψ.

Let A be the associated stiffness matrix. Then (14.2) is equivalent to the complementary relations

$$
\begin{aligned}
Av &\geq b, \\
v &\geq \psi, \\
(Av - b)'(v - \psi) &= 0.
\end{aligned}
$$

Here, we understand that $\psi_i = -\infty$ if there is no restriction at the point x_i.

Therefore the solution is a fixed point of the projected symmetrical Gauss–Seidel relaxation.

Projected Gauss–Seidel relaxation $S_h(v)$

for $i = 1, 2, \ldots, \dim(A)$ and for $i = \dim(A), \ldots, 2, 1$

$$
v_i \leftarrow \max\left\{\psi_i,\, v_i + \frac{\omega}{a_{ii}}(b_i - \sum_j a_{ij}v_j)\right\}. \tag{14.4}
$$

Here, it is understood that on the right-hand side of (14.4) always the current value of v_j is inserted. Of course, the relaxation can be applied with a relaxation factor ω between 0 and 2.

The computing time for m sweeps of the symmetric relaxation is roughly the same as for $m + 1/2$ matrix-vector multiplications. When evaluating (14.4), we can use the sum with the terms $i < j$ from the previous evaluation in the backward sweep and those with $i > j$ in the forward sweep.

In each relaxation step the quadratic functional given in (14.1) is reduced. Therefore, a compactness argument yields that the iteration with the projected Gauss-Seidel relaxation (as a one-level algorithm) converges to the solution [16]. It is however also known that it converges as slowly as the standard Gauss-Seidel relaxation for linear elliptic problems.

14.3   The Multilevel Procedure

For the setup of a multigrid algorithm we assume that we have a sequence of mesh sizes

$$
h_0 > h_1 > \ldots > h_L \tag{14.5}
$$

and $h_{\ell-1} = 2h_\ell$ for $\ell = 0, 1, \ldots, L$. The associated finite element spaces are assumed to be nested

$$V_{h_0} \subset V_{h_1} \subset \ldots \subset V_{h_L}, \tag{14.6}$$

and often we will write V_ℓ instead of V_{h_ℓ}.

As already mentioned, an essential feature of the cascadic algorithm is that the number m_ℓ of smoothing steps at the level ℓ increases for small ℓ. In the case of full regularity, it is sufficient to have $m_{\ell-1} = (2 + \varepsilon)m_\ell$; cf. [15].

Here we will be conservative for good reasons and choose

$$m_\ell = 3^{L-\ell} m_L. \tag{14.7}$$

Now we are ready to define our basic multilevel algorithm. Some improvements of the algorithm will be added later.

Cascadic Multigrid Algorithm.
1) Compute a good approximation v_0 of the solution for the coarsest level $\ell = 0$. [Here v_0 is considered as good if the residue is not larger than the expected residue for the final numerical solution on the highest level $(\ell = L)$.] — It can be computed by $S_h(v)$ with a suitable overrelaxation factor ω.

2) for $\ell = 1, 2, \ldots, L$

> {
>
> prolongate $v_{\ell-1} \rightarrow v_\ell$,
> apply $m_\ell = 3^{L-\ell} m_L$ steps of S_h
> improve the current solution
> > on the set of inactive points.
>
> }

The smoothing procedure in the algorithm above may consist only of the relaxations as described before. An improvement by cg steps will be described below. In the test examples below, suitable values for ω for the solution on the coarsest level and for the smoothing step are determined experimentally and are chosen as $\omega = 1.5$ and $\omega \approx 0.8$ respectively.

14.4   A Conjugate Gradient Algorithm

The cascadic algorithms for linear problems usually have conjugate gradient iterations in their smoothing procedures [15, 22, 63]. Moreover, a cg-iteration can accelerate the computation on the coarsest level. For these purposes we suggest a cg-method whose efficiency will become apparent from the numerical results.

Although our algorithm differs from the algorithm of Dostál and Schöberl [25], its concept is based on similar ideas:

1. The acceleration by steps of cg-type is restricted to the inactive set since such a restriction makes the problem linear. — In the case that restrictions are never active, the computation becomes the same as in the linear case.

2. When a computation differs for the active and the inactive set, the distinction between the two categories cannot be based on the question whether $v_i = \psi_i$ or $v_i > \psi_i$. Points with $v_i > \psi_i$ and $v_i - \psi_i$ being small must be treated like points of the active set.

3. Although the complete history of a cg-iteration enters into its analysis, the information of the last two steps is sufficient. The current step produces the optimum in the two-dimensional affine space that is parallel to the span of the current preconditioned gradient and the correction from the preceding step.

For convenience, we write the cg-procedure for the side conditions

$$u_i \geq 0 \qquad \text{for all} \quad i.$$

The generalization to $u_i \geq \psi_i$ (for some i) is obvious. Since we are discussing a one-level procedure, there is no specification of a mesh size.

Conjugate Gradient Algorithm for the Obstacle Problem (cg-PSSOR).

Let u^0 be an initial guess.

Compute u^1 by applying one step of the restricted symmetrical Gauss–Seidel relaxation $u^1 \leftarrow S(u^0)$, $d^0 = u^1 - u^0$

for $\nu = 1, 2, 3, \ldots$,

> {
> $\tilde{u}^{\nu+1} \leftarrow S(u^\nu)$,
> $g^\nu \leftarrow \tilde{u}^{\nu+1} - u^\nu$,
>   for $= 1, 2, \ldots, \dim(A)$
>   {
> $d_i^{\nu-1} \leftarrow g_i^\nu \leftarrow 0$ if $\max\{|d_i^{\nu-1}|, |g_i^\nu|\} \geq \tilde{u}_i^{\nu+1}$,
>   }
>   determine d^ν such that $\tilde{u}^{\nu+1} + d^\nu$ minimizes the
>   quadratic function on the set $u^{\nu+1} +$ span $\{d^{\nu-1}, g^\nu\}$,
> $u^{\nu+1} \leftarrow \tilde{u}^{\nu+1} + d^\nu$,
> }

14.5 Numerical Results

14.5.1 Example 1

As a first test example, we consider (14.2) on $\Omega := (0, 1)^2$ with $f = 0$. The obstacle is given by

$$\psi(x) := \exp\left((y^2 - 1)^{-1}\right),$$

with $y = (x - x_m)/r$, $x_m = (0.5, 0.5)$, $r = 0.4$.

In Table 14.1, we compare results with the projected Gauss-Seidel iteration (PSSOR) and with its conjugate gradient version (cg-PSSOR) when they are applied as one-level iterations. For simplicity, the relaxation parameter is kept fixed for all meshes. One observes the cg version to be significantly superior to the simple iteration. Reducing the mesh size by a factor of two by global refinement leads to only a doubling of the iteration steps. This indicates that the condition number κ of the problem enters with $\mathcal{O}(\sqrt{\kappa})$, as it is expected from the linear case.

| | cg-PSSOR ($\omega = 1.5$) | | PSSOR ($\omega = 1.5$) | |
DOF	iter.	res.	iter.	res.
1089	20	4.432-05	139	5.124-05
4225	36	7.488-05	448	8.335-05
16641	59	1.433-04	1439	1.647-04
66049	93	3.223-04	4393	3.276-04
263169	197	6.399-04	12106	6.497-04

Table 14.1: Example 1: Comparison of PSSOR and cg-PSSOR as one-level methods. The finite element spaces have between 1089 and 263169 degrees of freedom.

| | cg-PSSOR ($\omega = 0.75$) | | |
DOF	iter.	start-res.	end-res.
1089	37	8.803-03	8.339-10
4225	81	4.228-03	7.943-10
16641	27	2.130-03	2.250-05
66049	9	9.874-04	2.192-05
263169	3	4.888-04	2.897-05

| | PSSOR ($\omega = 0.75$) | | | | |
DOF	iter.	start-res.	end-res.	iter.	end-res.
1089	37	8.803-03	8.339-10	37	8.339-10
4225	81	4.228-03	7.826-05	162	3.033-05
16641	27	2.131-03	9.708-05	54	5.313-05
66049	9	9.909-04	7.262-05	18	4.174-05
263169	3	4.926-04	6.262-05	6	3.562-05

Table 14.2: Example 1: Results for cascadic iteration, comparing the performance of cg-PSSOR and PSSOR as smoothing procedure. — In all cases the cg-version with $\omega = 1.5$ was used as a solver on the coarsest level.

In Table 14.2, we show the results for our cascadic iteration, comparing the performance of cg-PSSOR and PSSOR as smoothing procedures. The number of smoothing steps (iter.), the initial residual (start-res.) and the final residual (end-res.) are depicted for each level. On the coarsest level (1089 DOF), the solution is computed nearly exactly by cg-PSSOR with $\omega = 1.5$ for both smoothing procedures.

At the start of the iteration the residual, measured in the natural energy norm, reflects the difference of the solutions of the h-grid and the $2h$ grid. So a reduction by a factor of 10 will lead to a residual error that is about $1/5$ of the discretisation error.

We observe the PSSOR smoother to require about twice as many smoothing iterations to yield approximately the same accuracy as the cg-PSSOR version. — For a comparison of the computing time we refer to the end of the paper.

In Figure 14.1 the dependence of the cascadic iteration on the relaxation parameter ω is depicted, when the smoothing is performed with PSSOR. The dependency suggests to choose $\omega = 0.75$ in the following test calculations. Moreover, the value of ω smaller than 1 makes clear that the iterations here serve as smoothers and not as solvers.

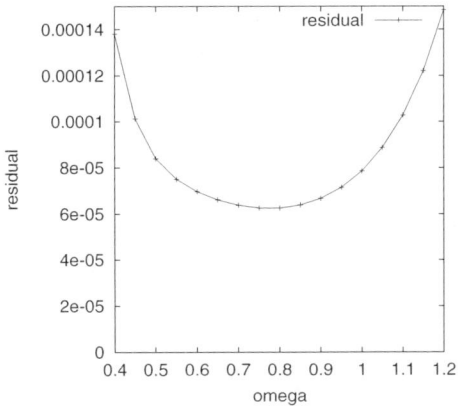

Figure 14.1: Example 1: Residual on final grid with 263169 degrees of freedom obtained by the cascadic iteration with relaxation parameter ω on each level.

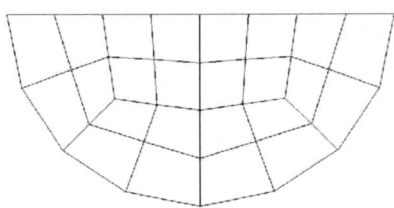

Figure 14.2: Coarse grid for Hertz contact problem

14.5.2 Example 2

Now we apply our ideas to Signorini's problem introduced in Chapter 6.

Following Kornhuber and Krause [46], we consider as a test example, a plane strain problem for a half circle with radius 0.4, centered at the point $(0; 0.4)$ in contact with a rigid plane (Figure 14.2). The material constants are given by $E = 270269N/mm^2$ for Young's modulus and $\nu = 0.248$ for Poisson's ratio. For our test calculations we choose a displacement of $(0; -0.05)$.

The results are presented the same way as for the example above.

Again, the superiority of the cg-PSSOR-method can be observed for a one level iteration; see Table 14.3. Within a cascadic procedure; see Table 14.4, again we observe the PSSOR smoother to require about twice as many smoothing iterations to reach approximately the same accuracy as the cg-PSSOR version.

We recall, that at the start of the iteration the residual, measured in the natural energy norm, is given by the difference of the solutions of the h-grid and the $2h$ grid. So a reduction by a factor of 10 will lead to a residual error that is about $1/5$ of the discretisation error.

Remark: Taking into account that one step of cg-PSSOR requires as many operations as two to three steps of PSSOR, the performance of both schemes can be regarded as equal for our examples. For more involved problems, for example anisotropic material behaviour, the cg-PSSOR is expected to be more robust as known from the linear case.

	cg-PSSOR ($\omega = 1.5$)		PSSOR ($\omega = 1.5$)	
DOF	iter.	res.	iter.	res.
834	47	6.476+00	543	8.265+00
3202	80	1.196+01	1651	1.204+01
12546	145	1.682+01	4971	1.726+01
49666	265	2.450+01	15123	2.459+01
197634	480	3.477+01	45221	3.489+01

Table 14.3: Example 2: Comparison between PSSOR and cg-PSSOR on one level, i.e., as solvers of the discrete problems.

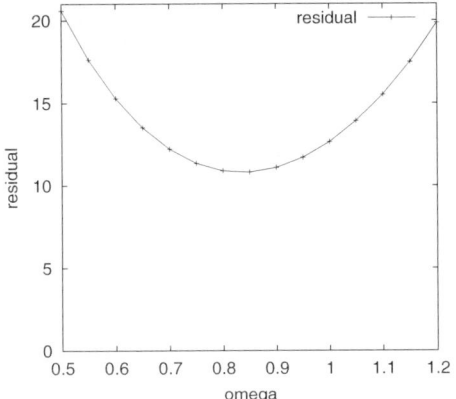

Figure 14.3: Example 2: Residual on final grid with 197634 degrees of freedom obtained by the cascadic iteration with relaxation parameter ω on each level.

		cg-PSSOR ($\omega = 0.85$)			
	DOF	iter.	start-res.	end-res.	
	834	97	1.111+03	9.965-10	
	3202	81	4.884+02	2.063-03	
	12546	27	2.377+02	1.031+00	
	49666	9	1.152+02	2.883+00	
	197634	3	5.692+01	4.171+00	
		PSSOR ($\omega = 0.85$)			
DOF	iter.	start-res.	end-res.	iter.	end-res.
834	97	1.111+03	9.965-10	97	9.965-10
3202	81	4.884+02	6.400+00	162	3.833+00
12546	27	2.379+02	7.466+00	54	4.005+00
49666	9	1.152+02	8.672+00	18	4.839+00
197634	3	5.688+01	1.077+01	6	4.897+00

Table 14.4: Example 2: Results for cascadic iteration, comparing the performance of cg-PSSOR and PSSOR as smoothing procedure.

Chapter 15

Conclusion

In this work, we have presented a general approach to error control and mesh adaptation (the DWR method) for finite element approximations of *variational inequalities*. By solving global dual problems *a posteriori* error estimates have been derived for the approximation of any prescribed quantity of physical interest. These error bounds can be evaluated by numerically solving (linearised) versions of the dual problems. The main emphasis has been on static problems of obstacle-type. Extensions to time dependent problems have been sketched.

The performance of the DWR method has been demonstrated for several model problems. In these tests the *dual-weighted* error estimators prove to be asymptotically sharp and provide the basis of constructing economical meshes. Effectivity comparisons are made with the traditional ZZ refinement indicator and partly with a heuristic global *energy-norm* estimator. The evaluation of the *weighted* error estimates requires on each mesh level to solve a (linearised) dual problem. In our experience, the extra work for mesh adaptation usually makes up less than 15 % of the total work on the *optimised* mesh.

The DWR method is a rather universal approach to adaptivity in Galerkin finite element schemes. For a discussion for the special case of (nonlinear) *variational equalities*, we refer to the surveys by Rannacher et al. [53, 9, 5] and the literature cited therin.

Appendix A

Algorithmic Aspects

The idea of the finite element method is quite universal since problems with different characteristic properties can be treated. As examples we mention the fields of reactive flow, radiative transfer and continuum mechanics. Therefore an approach to finite element code should reflect this universality. Most parts of the implementation of this method can be done problem independent. So it is advisable to use a program library, which for example supports the very complex task of grid-handling required for each special application. In 1991 Guido Kanschat and the author started at Bonn with the development of such library, namely DEAL (*Differential Equation Analysis Library*) (Becker, Kanschat & Suttmeier [21]).

In Kanschat [42] there are mentioned four development aims which are important in a complex software project, namely *computing speed, memory requirements, verifiability of code* and *flexibility*. We discuss the aspect of *flexibility* of our library at the example of the representation of a linear operator. DEAL offers the possibility to balance computing time and memory requirements with respect to the problem under consideration.

Due to our idea to represent an operator independent of boundary conditions and hanging nodes, these points have to be handled separately. At the end of this chapter we demonstrate our method of filtering techniques to take these points into account.

The finite element package

As mentioned above the universality is an important point in the finite element method. This means, first we look for properties all problems have in common. As an example we remark, that one main part of a finite element program consists of the geometric description of the domain and its discretisation in form of a triangulation. DEAL uses the object oriented concept for the grid-handling. A triangulation is regarded as an object, which consists of cells. These cells itself are described by their vertices.

We use a straightforward approach to organise the refinement process. In detail this is explained in Kanschat [42]. A triangle, for example, is divided into four congruent ones. The hierarchy is stored, i.e. all cells know their *father* and *children*. This enables us to do coarsening, which is a very important feature, especially in time dependent problems. As an example we refer to the sequence of grids at the end of Chapter 13. The moving transition zone between the elastic and the plastic part of the solution is resolved by the adaptive algorithm.

The complex problem of implementation of a triangulation is a good example to demonstrate the advantages of our approach. In a natural way the whole task is split into several small steps. These single steps are treated separately and therefore it is easy to improve the code locally. Furthermore, if it is necessary to change a concept we can reuse the single parts and we do not have to do changes within the whole library.

Representation of an operator

In the following we discuss some aspects of the treatment of partial differential equations to demonstrate that DEAL offers a high degree of *flexibility* without destroying the remaining three aims.

For simplicity, we assume that discretisation by the finite element method leads to a symmetric and positive definite operator A. The cg-method is one of the most frequently used solvers to treat such a linear system $Ax = b$. In figure A.1 we sketch parts of the template implementation of the cg-method used in DEAL. This solver requires only the abstract ability of the operator to do a matrix vector multiplication `vmult`. Therefore we recognise that the object oriented approach is adequate, since a `Triangulation` can be used as a basis for the further description of the problem class. Further derivations of this class include for example the different mechanisms to describe the operator of the special problem under consideration.

```
template<class MT, class VT, class MEM>
cg_simple(MT& A, VT& x, VT& b, ...)
{
  double res, alpha, beta;
  VT&     d, g, Ad;
  initialisation();
  for (int it=0;!reached;it++)
  {
    A.vmult(Ad,d);                // operator property

    alpha = Ad*d;    alpha = beta/alpha;
    x.add(alpha,d);  g.add(alpha,Ad);
    res = g*g;
    d.sadd(res/beta,-1.,g);
    beta = res;
    check_residual();
  }
}
```

Figure A.1: Pseudo code for template implementation of the cg-method

We sketch this with the two classes `Mechanics2D` and `Mechanics3D`. In the classical approach the operator describing a problem in continuum mechanics is represented by a matrix A. The memory requirements in the three dimensional case are very large. Therefore, it may be necessary to find an alternative of operator representation.

```
class Mechanics2D : Triangulation
{
    SparseMatrix A;
}

class Mechanics3D : Triangulation
```

```
{
    void vmult(Vector dst,Vector src);
}
```

In the following we discuss various possibilities to represent the operator. As a basis we choose the DEAL class `Triangulation`.

```
class Triangulation
{
    List of Cells;
};
```

All further implementation for the operator classes are derived from this class. For our purpose we mention the property, that a triangulation contains a list of cells. So the assembling process is for example done by

```
foreach active cell c
    c.assemble(...);
```

Classical matrix

The traditional approach to represent the operator is the use of a classical matrix. All information is saved in a global matrix. For example one can use a `SparseMatrix` where only non-zero entries are stored. So our classical approach leads to

```
class MatrixTriangulation : Triangulation
{
    SparseMatrix A;
    void vmult(Vector dst, Vector src);
};
```

The required matrix vector multiplication uses the already available properties of `SparseMatrix`. The method `A.vmult` is called indirectly by the `vmult` of the triangulation to have the possibility to incorporate for example filter routines described below. Essentially we have the structure

```
void MatrixTriangulation::vmult(Vector dst, Vector src)
{
   A.vmult(dst,src);

   ...

}
```

Cell matrices

Next, we want to describe a concept of storing the assemble information locally. Therefore we introduce the class LocalCell which contains the required memory in form of Matrix. Furthermore this cell is provided with a function vmult, which does the local multiplication.

```
class LocalCell : Cell
{
   Matrix A;
   void vmult(Vector dst, Vector src);
};

void LocalCell::vmult(Vector dst, Vector src)
{
   src -> local_src;
   A.vmult(local_dst,local_src);
   dst <- local_dst;
}
```

The required global matrix vector multiplication is now organised as a sum of the local vmult's.

```
class LocalTriangulation : Triangulation
{
   void vmult(Vector dst, Vector src);
};

void LocalTriangulation::vmult(Vector dst, Vector src)
{
   for each active cell c
```

```
      c.vmult(dst,src);
}
```

Scaled matrices

Most parts of the hierarchical structured triangulation consist of triangles or parallelograms. Consequently, the domain transformations are nearly all linear. Computing the relevant element matrices on the lowest level, the matrices required on the working level are computed simply by scaling the available knowledge from the lowest level. For more details about the scaling process itself see Kanschat [42].

```
class ScaleCell : Cell
{
    LowestLevelMatrix A;
    void vmult(Vector dst, Vector src);
};
```

With `LowestLevelMatrix` we indicate, that only memory allocation is required on the coarsest cells.

```
void ScaleCell::vmult(Vector dst, Vector src)
{
    A    -> local_A;
    src -> local_src;
    local_A.vmult(local_dst,local_src);
    dst <- local_dst;
}
```

Even in nonlinear problems one can use the scaling technique. For example in a primal finite element formulation for plasticity the nonlinearity is given in terms of ∇u_h. Treating this problem with a stationary iteration and using P_1-elements, the nonlinearity is constant over a cell. So the nonlinear term is simply a further scaling factor in addition to the domain scaling described above.

We can summarise, that already reliable and tested modules, for example the `SparseMatrix`, are used for the implementation. All work to implement the different operator classes is done locally and therefore global code like the the cg-method remains unchanged. So the point *verifiability of code* is fulfilled. If there are errors in the own program, we only have to examine the very few new code lines we did ourselves.

Due to the high degree of flexibility, there is the possibility to balance the points *computing speed* and *memory requirements.* If one wants to use elaborated standard code, i.e. BLAS libraries, one can choose the matrix oriented approach. On the other hand we use the scaling technique to solve problems which can only be treated reducing the memory requirements drastically. During an adaptive process only parts of the whole domain are refined. Computation time is saved because the local assembling is done only on new cells and the element matrices are stored locally.

Multigrid

Since the hierarchical structure of the triangulation is available, it suggests itself to use a multigrid algorithm in the context of solving algebraic equations resulting from the finite element method. The implementation was mainly done by G.Kanschat and R.Becker and is explained in detail in Becker [7]. Up to now the implementation requires the ability of cells to do vmult locally. So the used cells are based on the structures of LocalCell and ScaleCell. The user defined operator classes are derived from

```
class MGTriangulation : Triangulation
{
    hierarchical structure;
};
```

For problems arising in the present work our experience is, that local multigrid is very successful if a fixed number of iterations (say one or two) is used as a preconditioner in the cg-method. Using the cell wise approach sketched above one loses the possibility of matrix oriented smoothers like for example ILU-decomposition. But this is not so important since only spectral equivalence is needed between the operator A and the preconditioner. So for our purpose it is convenient to choose Gauss-Seidel-steps as a smoother. In addition we implemented another good alternative, namely doing a fixed number of cg-steps as a local smoother on each level of the triangulation. This concept only requires the operator property too. Furthermore, we want to remark that during an adaptive process in each step we have good initial guesses for the solution, coming from computations on grids of the previous adaptive step.

Filtering techniques and hanging nodes

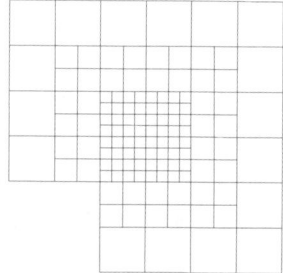
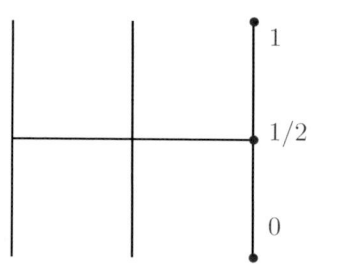

Figure A.2: Domain and sample mesh with hanging nodes

Figure A.3: Sketch of the appearance of a hanging node 1/2 depending on the nodes 0 and 1 for bilinear finite elements.

As mentioned before, for ease of mesh refinement and coarsening hanging nodes are allowed in our implementation, as shown in Figure A.2. The treatment of non-matching cells in the case of bilinear shape functions is sketched in Figure A.3.

Considerations in the following are restricted to Poisson's equation. Let $\{\varphi_1, \ldots, \varphi_N\}$ denote the standard nodal basis for V_h with dimension N. We define the vectors of nodal values $U, F \in \mathbb{R}^N$ with

$$U_i = (\nabla u_h, \nabla \varphi_i), \quad F_i = (f, \varphi_i),$$

and the corresponding matrix $A \in \mathbb{R}^{N \times N}$ with $A_{ij} = (\nabla \varphi_j, \nabla \varphi_i)$.

Since continuity of the solution is required, u_h has to be linear along $\overline{01}$, i.e. assuming u_h determined by U_0 and U_1 in nodes 0 and 1, the value for u_h in node 1/2 has to be subjected to the constraint $U_{1/2} = 0.5(U_0 + U_1)$.

Then introducing the averaging operator M the constraints for ensuring continuity of u_h can be written in the form $U = MU$. The operator M acts as an identity on regular points 0 and 1 and performs an averaging on hanging nodes 1/2 giving the structure

$$M = \begin{pmatrix} 1 & & & & \\ & \ddots & & & \\ & \frac{1}{2} & 0 & \frac{1}{2} & \\ & & & \ddots & \\ & & & & 1 \end{pmatrix}.$$

Consequently one has to solve $AU = F$ with the additional constraint $U = MU$. Possibilities of incorporating the constraint into FE-implementations are discussed separately in a section below.

Elements of higher order and nonconforming discretisations

The techniques described above for the case of bilinear shape functions can be extended to other situations. We discuss here the biquadratic and the nonconforming Crouzeix-Raviart element.

Biquadratic

For the biquadratic shape functions, let the vector on the edge with the hanging nodes be arranged in the form

$$U^T = \begin{pmatrix} U_0 & U_{1/4} & U_{1/2} & U_{3/4} & U_1 \end{pmatrix}$$

(see Figure A.4, right). In order to preserve the continuity, u_h has to be quadratic along $\overline{01}$, u_h determined by U_0, $U_{1/2}$ and U_1. Hence

$$U_{1/4} = \frac{3}{8}U_0 + \frac{3}{4}U_{1/2} - \frac{1}{8}U_1$$

$$U_{3/4} = -\frac{1}{8}U_0 + \frac{3}{4}U_{1/2} + \frac{3}{8}U_1$$

and the average operator is given by

$$
M = \begin{pmatrix}
1 & & & & \\
3/8 & 0 & 3/4 & 0 & -1/8 \\
& & 1 & & \\
-1/8 & 0 & 3/4 & 0 & 3/8 \\
& & & & 1
\end{pmatrix}.
$$

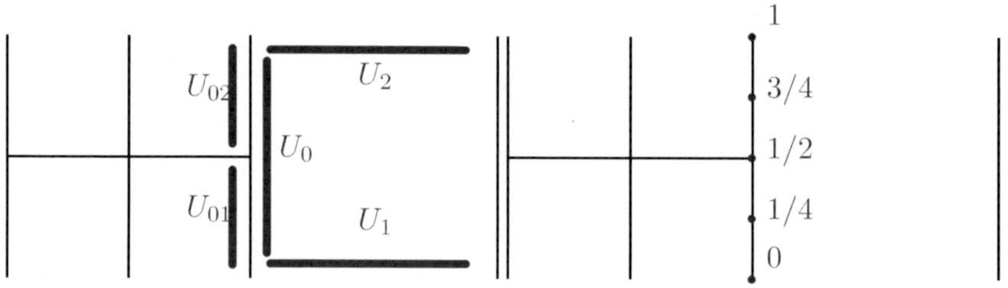

Figure A.4: Hanging nodes for rotated bilinear (left) and biquadratic (right) finite elements. Bars or points indicate degrees of freedom.

Rotated bilinear

Eventually, we discuss the extension of the Crouzeix-Raviart element on triangles to quadrilaterals, i.e., the rotated bilinear element $\tilde{Q}_1$ analyzed by Rannacher and Turek in [59]. Analogously to the triangular elements, we take the mean value along each edge as node values. The required restrictions to the degrees of freedom in situations sketched in Figure A.4 are imposed as follows. Remarking, that the node values of the rotated bilinear element are related to the element edges and are in fact the mean value along these lines, conformity with respect to the chosen discrete space is ensured by

$$
U_{01} = U_0 + \frac{1}{4}U_1 - \frac{1}{4}U_2
$$
$$
U_{02} = U_0 - \frac{1}{4}U_1 + \frac{1}{4}U_2 \,.
$$

Implementation

In traditional approaches the treatment of Dirichlet data can be done by eliminating rows and columns of the matrix. We propose to use filtering techniques so that the representation of the operator is independent of the desired boundary values. We assume that the operator is implemented in form of a class Matrix with a default matrix-vector multiplication vmult.

The used cg-like methods are based on an update step for the solution x of the form $x_{k+1} = x_k + \alpha_k d_k$, where d_k is a certain correction vector. So after initialising the iteration vector with the correct Dirichlet values, we have to ensure that they are not changed during the iteration. This is simply done by modifying the matrix vector multiplication. With terms introduced above, all entries in the correction vector d_k, representing Dirichlet points, are set to zero after the vmult of the operator has taken place.

```
void FilterMatrix::vmult(Vector dst, Vector src)
{
    A.vmult(dst,src);          // default vmult
    set_zero_boundary(dst);
}
```

Carrying over this idea to the treatment of hanging nodes, we rewrite the discrete Poisson equation as a minimum problem

$$U^T A U - 2U^T F = \min,$$
$$U = MU,$$

or in compact form

$$U^T (M^T A M) U - 2U^T M^T F = \min.$$

We see that the appropriate modification of the matrix vector multiplication consists simply of the two processes M and M^T referred to as average and distribute respectively. So in our pseudo programming language the modified vmult-version looks like

```
void FilterMatrix::vmult(Vector dst, Vector src)
{
    average(src);
    A.vmult(dst,src);    // default vmult
    distribute(dst);
    set_zero_boundary(dst);
}
```

Bibliography

[1] M. Ainsworth and J.T. Oden. A posteriori error estimation in finite element analysis. *Comput. Meth. Appl. Mech. engrg.*, 142:1–88, 1997.

[2] M. Ainsworth, J.T. Oden, and C.Y. Lee. Local a posteriori error estimators for variational inequalities. *Numerical Methods for Partial Differential Equations*, 9:23–33, 1993.

[3] I. Babuška and A.D. Miller. A feedback finite element method with a posteriori error estimation, part 1. *Comput. Meth. Appl. Mech. Engrg.*, 61:1–40, 1987.

[4] E. Backes. Gewichtete a posteriori Fehleranalyse bei der adaptiven Finite-Elemente-Methode: Ein Vergleich zwischen Residuen- und Bank-Weiser-Schätzer. Master's thesis, Institut für Angewandte Mathematik, Universität Heidelberg, 1997.

[5] W. Bangerth and R. Rannacher. *Adaptive finite element methods for differential equations.* Birkhäuser Verlag, Basel, Boston, Berlin, 2003.

[6] R.E. Bank and A. Weiser. Some a posteriori error estimators for elliptic partial differential equations. *Math. Comp.*, 44:283–301, 1985.

[7] R. Becker. An Adaptive Finite Element Method for the Incompressible Navier-Stokes Equations on Time-Dependent Domains. *Dissertation, Institut für Angewandte Mathematik, Universität Heidelberg*, 1995.

[8] R. Becker and R. Rannacher. A feed-back approach to error control in finite element methods: Basic analysis and examples. *EAST-WEST J. Numer. Math.*, 4:237–264, 1996.

[9] R. Becker and R. Rannacher. An optimal control approach to a-posteriori error estimation. in a. iserles, editor. *Acta Numerica, Cambridge University Press*, pages 1–102, 2001.

[10] A. Bensoussan and J. Frehse. Asymptotic behaviour of the time dependent norton-hoff law in plasticity theory an h^1 regularity. *Comment. Math. Univ. Carilinae*, 37(2), 1996.

[11] H. Blum, D. Braess, and F.T. Suttmeier. A cascadic multigrid algorithm for variational inequalities. *Computing and Visualization in Science*, 7:153–157, 2004.

[12] H. Blum and F.T. Suttmeier. An adaptive finite element discretisation for a simplified Signorini problem. *Calcolo*, 37(2):65–77, 2000.

[13] H. Blum and F.T. Suttmeier. Weighted Error Estimates for Finite Element Solutions of Variational Inequalities. *Computing*, 65:119–134, 2000.

[14] C. Bollrath. Two multi-level algorithms for the dam problem. In D. Braess, W. Hackbusch, and U. Trottenberg, editors, *Advances in multigrid methods*. Vieweg & Sohn, Braunschweig/Wiesbaden, 1984.

[15] F.A. Bornemann and P. Deuflhard. The cascadic multigrid method for elliptic problems. *Numer. Math.*, 75:135–152, 1996.

[16] A. Brandt and C.W. Cryer. Multigrid algorithms for the solution of linear complementarity problems arising from free boundary problems. *SIAM J. Sci. Stat. Comput.*, 4:655–684, 1980.

[17] H. Brézis. Problèmes unilatéraux. *(Thèse) J.Math. Pures Appl.*, 1972.

[18] H. Brézis and G. Stampacchia. Sur la régularité de la solution d'inéquations elliptiques. *Bull. Soc. Math. France*, 96:153–180, 1968.

[19] F. Brezzi, W.W. Hager, and P.A. Raviart. Error estimates for the finite element solution of variational inequalities. *Num. Math.*, 28:431–443, 1977.

[20] C. Carstensen, O. Scherf, and P. Wriggers. Adaptive finite elements for elastic bodies in contact. *SIAM J. Sci. Comput.*, 20(5):1605–1626, 1999.

[21] DEAL. differential equations analysis library
http://www.math.uni-siegen.de/suttmeier/deal/deal.html, 1995.

[22] P. Deuflhard. Cascadic conjugate gradient methods for elliptic partial differential equations: algorithm and numerical results. In D. Keyes and J. Xu, editors, *Domain Decomposition Methods in Scientific and Engineering Computing*, volume 180 of *AMS Series*, pages 29–42, 1994.

[23] M. Dobrowolski and T. Staib. On finite element approximation of a second order unilateral variational inequality. *Numer. Funct. Anal. and Optimiz.*, 13:243–247, 1992.

[24] W. Dörfler and M. Rumpf. An adaptive strategy for elliptic problems including a posteriori controlled boundary approximation. *Math. Comp.*, 67(224):1361–1382, 1998.

[25] Z. Dostál and J. Schöberl. Minimizing quadratic functions subject to bound constraints. Technical report, Tech. Univers. Ostrava, 2002. in preparation.

[26] G. Duvaut and J. L. Lions. *Inequalities in Mechanics and Physics*. Springer, Berlin-Heidelberg-New York, 1976.

[27] B. Erdmann, M. Frei, R.H.W. Hoppe, R. Kornhuber, and U. Wiest. Adaptice finite element methods for variational inequalities. *East-West J. Numer. Math.*, 1(3):165–197, 1993.

[28] B. Erdmann, R.H.W. Hoppe, and R. Kornhuber. Adaptive multilevel-methods for obstacle problems in three space dimensions. In W. Hackbusch et al, editor, *Adaptive methods - algorithms, theory and applications*, volume 46 of *Notes Numer. Fluid Mech.*, pages 20–141. Vieweg, 1994.

[29] K. Eriksson, D. Estep, P. Hansbo, and C. Johnson. Introduction to adaptive methods for partial differential equations. *Acta Numerica*, 4:105–159, 1995.

[30] R.S. Falk. Error estimates for the approximation of a class of variational inequalities. *Math. Comp.*, 28:963–971, 1974.

[31] S. Falk and B. Mercier. Error estimates for elasto-plastic problems. *R.A.I.R.O. Analyse Numérique/Numerical Analysis*, 11, $n°$ 2:135–144, 1977.

[32] L.P. Franca and R. Stenberg. Error analysis of some Galerkin least squares methods for the elasticity equations. *SIAM J. Numer. Anal.*, 28, 1991.

[33] R. Glowinski. *Numerical methods for nonlinear variational problems.* Springer Series in Comp. Physics. Springer, 1983.

[34] R. Glowinski, J.L. Lions, and R. Trémolières. *Numerical Analysis of Variational Inequalities.* North-Holland, 1981.

[35] W. Hackbusch and H.D. Mittelmann. On multi-grid methods for variational inequalities. *Numer. Math.*, 42:65–76, 1983.

[36] P. Hansbo and C. Johnson. Adaptive finite element methods for elastostatic contact problems. *in 'Grid Generation and Adaptive Algorithms', Editors: Marshall Bern, Joseph E. Flaherty, and Mitchell Luskin, Springer*, 1999.

[37] R.H.W. Hoppe and R. Kornhuber. Adaptive multilevel methods for obstacle problems. *SIAM J. Numer. Anal.*, 31(2):301–323, 1994.

[38] C. Johnson. On finite element methods for plasticity problems. *Numer. Math.*, 26:79–84, 1975.

[39] C. Johnson. A mixed finite element method for plasticity problems with hardening. *SIAM J.Numer. Anal.*, 14, 1977.

[40] C. Johnson. On Plasticity with Hardening. *Journal of math. Analysis and Applications*, 62:325–336, 1978.

[41] C. Johnson and P. Hansbo. *Adaptive finite elemente methods in computational mechanics.* Computer Methods in Applied Mechanics and Engineering 101. North-Holland, 1992.

[42] G. Kanschat. Parallel and Adaptive Galerkin Methods for Radiative Transfer Problems. *Dissertation, Institute for Applied Mathematics, University of Heidelberg*, 1996.

[43] N. Kikuchi and J.T. Oden. *Contact Problems in Elasticity: A Study of Variational Inequalities and Finite Element Methods.* Studies in Applied Mathematics 8. SIAM, 1988.

[44] D. Kinderlehrer and G. Stampacchia. *An introduction to variational inequalities and their applications.* Academic Press, 1980.

[45] R. Kornhuber. A posteriori error estimates for elliptic variational inequalities. *Computers Math. Applic.*, 31:49–60, 1996.

[46] R. Kornhuber and R. Krause. Adaptive multigrid methods for signorini's problem in linear elasticity. *Comp. Visual. Sci.*, 4:9–20, 2001.

[47] P. Ladevéze and D. Leguillon. Error estimate procedure in the finite element method and applications. *SIAM J. Numer. Anal.*, 20:485–509, 1983.

[48] J.L. Lions and G. Stampacchia. Variational Inequalities. *Comm. Pure Appl. Math.*, 20:493–519, 1967.

[49] C. Großmann and H.-G.Roos. *Numerik partieller Differentialgleichungen.* Teubner Studienbücher, 1992.

[50] U. Mosco. Error estimates for some variational inequalities. In E.Magenes I.Galligani, editor, *Mathematical Aspects of Finite Element Methods*, volume 606 of *Lecture Notes in Math.*, pages 224–236. Springer-Verlag, 1977.

[51] F. Natterer. Optimale L^2-Konvergenz Finiter Elemente bei Variationsungleichungen. *Bonner Math. Schr.*, 89:1–12, 1976.

[52] J. Nitsche. l^∞-convergence of finite element approximations. In E.Magenes I.Galligani, editor, *Mathematical Aspects of Finite Element Methods*, volume 606 of *Lecture Notes in Math.*, pages 261–274. Springer-Verlag, 1977.

[53] R. Rannacher. *Error control in finite element computations.* Proc. NATO-Summer School *Error Control and Adaptivity in Scientific Computing*, Antalya, Turkey, Aug. 9-12. Kluwer-Verlag, 1998.

[54] R. Rannacher. A posteriori error estimation in least-squares stabilized finite element schemes. *Comp. Meth. in Appl. Mech. and Eng.*, to appear. Special issue on Advances in Stabilized Methods in Computational Mechanics.

[55] R. Rannacher and F.T. Suttmeier. A feed-back approach to error control in finite element methods: Application to linear elasticity. *Comp. Mech.*, 19(5):434–446, 1997.

[56] R. Rannacher and F.T. Suttmeier. A posteriori error control in finite element methods via duality techniques: application to perfect plasticity. *Comp. Mech.*, 21(2):123–133, 1998.

[57] R. Rannacher and F.T. Suttmeier. A posteriori error estimation and mesh adaptation for finite element models in elasto-plasticity. *Comput. Meth. Appl. Mech. Engrg.*, 176:333–361, 1999.

[58] R. Rannacher and F.T. Suttmeier. Error estimation and adaptive mesh design for FE models in elasto-plasticity. In E. Stein, editor, *Error-Controlled Adaptive FEMs in Solid Mechanics*. John Wiley, 2002.

[59] R. Rannacher and S. Turek. Simple nonconforming quadrilateral stokes element. *Num. Meth. PDE*, pages 97–111, 1992.

[60] G.A. Seregin. On the regularity of weak solutions of variational problems in plasticity theory. *Soviet Math. Dokl.*, 42(2), 1991.

[61] R. Seydel. *Tools for Computational Finance*. Springer, 2002.

[62] F.S. Shaw. The torsion of solid and hollow prisms in the elastic and plastic range by relaxation methods. Technical report, ACA-11, 1944.

[63] R. Stevenson. Nonconforming finite elements and the cascadic iteration. *Numer. Math.*, 91:351–387, 2002.

[64] G. Strang. *A minimax problem in plasticity theory*. Functional analysis methods in numerical analysis, Spec. Sess., AMS, St. Louis 1977, Lect. Notes Math. 701. 319-333. Springer, 1979.

[65] P.-M. Suquet. *Existence et Rgularit de Solutions des Equations de la Plasticit parfaite*. PhD thesis, Universite Paris Sud, 1978. These de 3me cycle.

[66] F.T. Suttmeier. Adaptive Finite Element Approximation of Problems in Elasto-Plasticity Theory. *Dissertation, Institute for Applied Mathematics, University of Heidelberg*, 1996.

[67] F.T. Suttmeier. On FE-discretisations with least-squares stabilisation: A posteriori error control for the membrane-problem. *EAST-WEST J. Numer. Math.*, 6:155–165, 1998.

[68] F.T. Suttmeier. Error analysis for finite element solutions of variational inequalities. Professorial dissertation, University of Dortmund, July 2001. http://www-lsx.mathematik.uni-dortmund.de/lsx/paper/habil.ps.gz.

[69] F.T. Suttmeier. General approach for a posteriori error estimates for finite element solutions of variational inequalities. *Computational Mechanics*, 27(4):317–323, 2001.

[70] L.P. Franca T.J.R. Hughes and M. Balestra. *A new finite element formulation for computational fluid dynamics: V. Circumventing the Babuška-Brezzi condition: A stable Petrov-Galerkin formulation for the Stokes problem accommodating equal order interpolation. Comp. Meth. Appl. Mech. Eng.*, 59:85–99, 1986.

[71] A. Veeser. Efficient and reliable a posteriori error estimators for elliptic obstacle problems. *SIAM J. Numer. Anal.*, to appear.

[72] R. Verfürth. *A Review of a posteriori Error Estimation and Adaptive Mesh-Refinement Techniques*. B.G. Wiley-Teubner, Stuttgart, 1996.

[73] C. Wieners. Robust multigrid methods for nearly incompressible elasticity. *Computing*, 64(4):289–306, 2000.

[74] O.C. Zienkiewicz and J.Z. Zhu. A simple error estimator and adaptive procedure for practical engineering analysis. *Int. J. Numer. Methods Engrg.*, 24:337–357, 1987.